T0140333

Studies in Systems, Decision and Control

Volume 274

Series Editor

Janusz Kacprzyk, Systems Research Institute, Polish Academy of Sciences, Warsaw, Poland

The series "Studies in Systems, Decision and Control" (SSDC) covers both new developments and advances, as well as the state of the art, in the various areas of broadly perceived systems, decision making and control–quickly, up to date and with a high quality. The intent is to cover the theory, applications, and perspectives on the state of the art and future developments relevant to systems, decision making, control, complex processes and related areas, as embedded in the fields of engineering, computer science, physics, economics, social and life sciences, as well as the paradigms and methodologies behind them. The series contains monographs, textbooks, lecture notes and edited volumes in systems, decision making and control spanning the areas of Cyber-Physical Systems, Autonomous Systems, Sensor Networks, Control Systems, Energy Systems, Automotive Systems, Biological Systems, Vehicular Networking and Connected Vehicles, Aerospace Systems, Automation, Manufacturing, Smart Grids, Nonlinear Systems, Power Systems, Robotics, Social Systems, Economic Systems and other. Of particular value to both the contributors and the readership are the short publication timeframe and the world-wide distribution and exposure which enable both a wide and rapid dissemination of research output.

Indexed by SCOPUS, DBLP, WTI Frankfurt eG, zbMATH, SCImago.

More information about this series at http://www.springer.com/series/13304

Fiedelholtz

The Cyber Security Network Guide

 Springer

Fiedelholtz
UMBC
Hilltop Circle, Baltimore, MD, USA

ISSN 2198-4182 ISSN 2198-4190 (electronic)
Studies in Systems, Decision and Control
ISBN 978-3-030-61593-2 ISBN 978-3-030-61591-8 (eBook)
https://doi.org/10.1007/978-3-030-61591-8

This Springer imprint is published by the registered company Springer Nature Switzerland AG
The registered company address is: Gewerbestrasse 11, 6330 Cham, Switzerland

Preface

The Cyber Network Guide, *Studies in Systems, Decision and Control* textbook provides a cybersecurity technical foundational landscape of the international government cyber operational capabilities for college undergraduate and graduate students, which includes standard operating procedures, cyber operational documentation, and description of tools utilized in the event of a major cyber incident.

The cyber textbook provides in-depth background information for each step of the cyber analytic framework process, which includes the following requirements:

- Pre-incident Planning and Analysis
- Incident Detection and Characterization
- Vulnerability/Consequence Analysis
- Incident Response and Recovery
- Cloud Architecture Analysis

The purpose of the Cyber Network Guide, Studies in Systems, Decision and Control book is to standardize requirements for cyber-physical international operational analysts and response for cyber practitioners.

This unique technical guide provides cyber analysts with appropriate *Open Source* background information to underpin efforts to provide accurate and comprehensive inputs in the development of cyber analytic products. In addition, this cyber primer is an effort to provide cyber baseline information in collaboration with international governments regarding vulnerability and consequence analysis for responding to cyber network attacks.

Baltimore, USA   Fiedelholtz

Acknowledgements

I would like to dedicate this cyber book to Jennifer Fiedelholtz, who has stood by me with all my crazy career changes and who gave me two beautiful children, Sarah and Matthew.

In addition, this Cyber Network book would not be possible if not for the tireless efforts of the team at *Media Graphics Lab*, Rockville, Maryland. The excellent graphic diagrams in this book are a reflection of their graphics skills and command of the computer science principles encapsulated in this material.

Finally, while teaching at the University of Maryland at Baltimore (UMBC), I noticed that a more hands-on approach to supplement their excellent computer science pedagogical foundational approach is needed. This cyber book will address that gap to provide the university with both a computer science comprehensive theoretical and hands-on skills.

Baltimore, USA   Fiedelholtz

Introduction

Chapter 1 Pre-incident Planning and Analysis

The purpose of cyber-physical planning and analysis under the Cyber Network Guide, Studies in Systems, Decision and Control textbook is to standardize requirements for cyber-physical preparedness (**including vulnerability and consequence analysis**) and operational incident response for all international cyber stakeholders.

Chapter 2 Incident Detection and Characterization

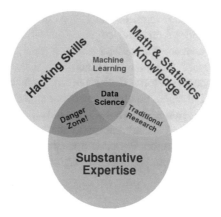

The Cyber Operation Centers will continue to serve as a centralized location where operational elements involved in cybersecurity and communications dependence are coordinated and integrated. The cyber incident response partners include all Federal departments and agencies; State, local, tribal, and territorial governments; the private sector; and international partners. In close coordination with the originators of information and with other partners, the Federal requests, receives, shares, and analyzes information on cyber-attack techniques and vulnerabilities from the range of the highest to the lowest level of classification or restriction possible and works with Federal partners to mitigate threats to critical networks.

In this process, the Federal government continues as the information and coordination hub of a national network to protect critical infrastructure. Specific Federal government roles include situational awareness and crisis monitoring of critical infrastructure, and information sharing on threat information and collaboration, assessment and analysis, and decision support pre- and post-incident.

Both the Federal government and private sector will assist and collaborate sector analysts and provide actionable information in real time for comprehensive cyber and physical analysis of critical infrastructure during an incident (manmade and/or natural disaster). The centers' activities include analysis and providing a greater situational awareness of cybersecurity and communications vulnerabilities, intrusions, incidents, mitigation, and recovery actions.

Chapter 3 Vulnerability/Consequence Analysis

Section 3.1 Information Sharing

Secure, functioning, and resilient critical networks require the efficient exchange of information, including intelligence, between all levels of government and critical infrastructure sector owners and operators. This must facilitate the timely exchange of threat and vulnerability information as well as information that allows for the development of a situational awareness capability during incidents.

To conduct vulnerability/consequence analysis of any particular sector or potential exploit, the analyst will collect relevant data from cyber operational centers on credible cyber threats and combine it with cyber-physical system information from owner/operators. Collection of data includes understanding potential cyber exploit details, potential impacts on the entity based on their information, and relevant mitigation steps. The analyst will leverage partnerships with the government, industry, and international partners to determine risk to infrastructure owners and operators from credible cyber threats. These analyses will provide the basis for pre-incident preparedness activities and post-incident response and recovery.

Chapter 4 Incident Response and Recovery

The cyber analyst in the cyber-physical response and recovery will provide necessary support in the analysis and development of response and recovery action plans to protect the sector entity or infrastructure property, business needs, and the environment after an incident has occurred and then continue this effort in stabilizing the incident. Response and recovery efforts are focused on ensuring that the sector entity or infrastructure is able to effectively respond to any threats with an emphasis on economy, environment, and safety.

The ensuing recovery process includes those capabilities necessary to assist the sector entity or infrastructure asset, as well as communities affected by an incident, in recovering effectively and prioritizing action plans and support. It is focused on timely restoration, strengthening, and revitalization of the infrastructure. Successful recovery requires informed and coordinated leadership, collaboration between the sector partners during all phases of the recovery process.

Chapter 5 Cloud Architecture

Cloud computing is currently an essential part of network infrastructure. The U.S. National Institute for Standards and Technology has proposed defining cloud computing as a model "for enabling convenient, on-demand network access to a shared pool of configurable computing resources." The cloud structure facilitates the user to access the real-time metadata in a quick and more efficient manner. Data uploads into a centralized data center and is distributed through the network in milliseconds.

Chapter 6 Lessons Learned

Once the cyber-physical analysis and incident response and recovery action and plans are communicated to the sectors or entities in question, the cyber operational agencies, sector agencies, and the sector or entity involved should review the entire incident analysis process to determine what worked and identify areas of weakness to improve the process for the next incident response and recovery analysis.

The lessons learned process should strive to identify shortcomings of the process that walks through the Cyber-Physical Mapping Framework Analysis Process Matrix depicted in Appendix G (i.e., pre-incident planning and analysis, incident detection and characterization, vulnerability/consequence analysis, incident response and recovery, roles, responsibilities, collaboration, support, various report outputs, and training). The lessons learned should provide guidance for the additional need for training for all the personnel involved in the cyber-physical incident analysis.

The lessons learned should also focus on collecting data, sharing information, providing guidance and assisting the sector or entity in preventing future attacks, preventing or limiting disruption if they do occur, and creating early visibility of such attacks through enhanced awareness, security monitoring, and training.

Contents

List of Figures

Chapter 1
Pre-incident Planning and Analysis

The purpose of cyber-physical planning and analysis under the Cyber Network Guide, Studies in Systems, Decision and Control textbook is to standardize requirements for cyber-physical preparedness (**including vulnerability and consequence analysis**) and operational incident response for all international cyber stakeholders.

1.1 Steady-State and Continuous Monitoring

The cyber and private sector international organizations under this cyber-physical incident process and monitoring will, in advance, have the mechanisms and facility to allow the development of a common operational picture of the incident in question based on participation and input. Preparation also includes training to ensure individuals, teams, and organization leadership are trained in cyber-incident response procedures and internal/external reporting mechanisms. Training should also ensure that individuals meet professional qualifications and performance standards and have been trained in their specific cyber roles within their organization structure and process.

During steady-state and monitoring activity, the international cyber organizations will leverage expertise in working with the international cyber governments and private sector organizations to identify critical areas for vulnerability and consequence analysis as it relates to infrastructure dependency, interdependency, cascading effects, and cross-sector impacts of a potential incident. Cyber analysts will collaborate to standardize understanding of roles and responsibilities among the partners (Fig. 1.1). Each organization plays a unique role in preparing for a cyber incident with respect to its distinct mission and organizational authorities.

In the context of the United States National Infrastructure Protection Plan (NIPP), "steady state" is defined as the posture for routine, normal, day-to-day operations, as contrasted with temporary periods of heightened alert or real-time response to

© Springer Nature Switzerland AG 2021
Fiedelholtz, *The Cyber Security Network Guide*, Studies in Systems,
Decision and Control 274, https://doi.org/10.1007/978-3-030-61591-8_1

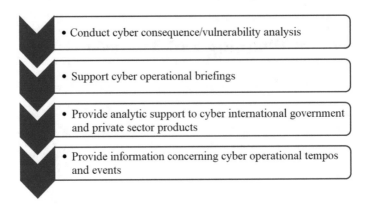

Fig. 1.1 Cyber incident standard operating procedure (SOP)

threats or incidents. Activities completed by the international cyber analysts during the steady-state period include the following:

For roles, responsibilities, and processes in the cyber-physical incident planning and response framework, the cyber analysts should refer to the most recent Gartner report referencing cyber-attacks and disruption of business activities.[1]

---

[1]"How to Prepare for and Response to Business Disruption After Aggressive Cyber Attacks", August 2019.

Chapter 2
Incident Detection and Characterization

The Cyber Operation Centers will continue to serve as a centralized location where operational elements involved in cybersecurity and communications dependence are coordinated and integrated. The cyber incident response partners include all Federal departments and agencies; State, local, tribal, and territorial governments; the private sector; and international partners. In close coordination with the originators of information and with other partners, the Federal requests, receives, shares, and analyzes information on cyber-attack techniques and vulnerabilities from the range of the highest to the lowest level of classification or restriction possible and works with Federal partners to mitigate threats to critical networks.

In this process, the Federal government continues as the information and coordination hub of a national network to protect critical infrastructure. Specific Federal government roles include situational awareness and crisis monitoring of critical infrastructure, and information sharing on threat information and collaboration, assessment and analysis, and decision support pre- and post-incident.

Both the Federal government and private sector will assist and collaborate sector analysts and provide actionable information in real time for comprehensive cyber and physical analysis of critical infrastructure during an incident (manmade and/or natural disaster). The centers' activities include analysis and providing a greater situational awareness of cybersecurity and communications vulnerabilities, intrusions, incidents, mitigation, and recovery actions.[1]

2.1 Detection

When prevention and protection efforts are unsuccessful, Federal, State, local, tribal, territorial, and private-sector owners and operators of critical networks and infrastructure are likely to be the first to detect malicious or unauthorized activity on their

---

[1]Federal [1].

© Springer Nature Switzerland AG 2021
Fiedelholtz, *The Cyber Security Network Guide*, Studies in Systems,
Decision and Control 274, https://doi.org/10.1007/978-3-030-61591-8_2

networks. In general, the critical infrastructure sector owners and operators work independently and within their company's incident response processes to address cybersecurity issues. These private-sector owners and operators in partnership with others, when appropriate, identify and contain malicious and unauthorized activity on their critical networks (from internal/external cyber attacks). They seek to gather as much information as possible on the unauthorized activity, including any critical details on the "who, what, where, when, why, and how" of the incident, if known. As part of this activity, organizations reporting an incident may report directly to the Federal Operation Centers through its collocated organizations; indirectly through the Situation Operation Center (SOC), National Operation Center (NOC), or law enforcement, intelligence, and regulatory agencies; indirectly through Information Sharing and Analysis Centers (ISACs); and/or directly from the private sector.[2] Activities related to detection conducted by the Cyber Analyst in collaboration with Federal Cyber Operation Centers may include the following:

- **Conduct open source monitoring**—Monitor the Cyber Daily Open Source Infrastructure Report, a summary of open-source published information collected each business day concerning significant critical cyber issues. Each Daily Report is divided by critical infrastructure sectors and key assets defined in the NIPP, and discusses relevant physical and cyber incidents across the Nation.[3]
- **Collect datasets (security incidents, alerts, and events)**—During the cyber incident, review and collect, major cyber reports, as well as sector reports of essential security incidents, alerts, events, and information on control systems in sector area(s) of concern. Such reports may include the Suspicious Activity Report (SAR), U.S. Computer Emergency Readiness Team (US-CERT), and Industrial Control System-Computer Emergency Readiness Team (ICS-CERT) weekly and monthly reports; Open Source Infrastructure report and analysis; and others.
- **Identify and analyze suspicious activity**—Identify and analyze suspicious activity reporting (SAR) from the Nationwide Suspicious Activity Reporting Initiative (NSI) database. The SAR initiative is a collaborative effort led by the U.S. Department of Justice (DOJ), Bureau of Justice Assistance, in partnership with the U.S. Department of Homeland Security (DHS), the Federal Bureau of Investigation (FBI), and State, local, tribal, and territorial law enforcement partners. The program establishes a national capacity for gathering, documenting, processing, analyzing, and sharing SAR information among law enforcement agencies. SAR reports are vetted by the cyber fusion centers and shared as appropriate among NSI participants.
- **Detect and verify unusual and network traffic**—Cyber Operation Centers, will review, detect, and verify real-time cyber data with automatic collection tools, and conduct deep packet inspection of traffic coming to or from Federal Internet protocol (IP) addresses ending in ".com" or ".gov" to detect signs of suspicious or malicious activities.

---

[2] Federal [2].
[3] Federal [3].

- **Review and collect sector reports of essential information systems**—Review and collect relevant critical infrastructure sector reports and information through existing partnership agreements with critical infrastructure owners/operators, Federal agencies, and State/local governments:

  - **Critical infrastructure owners/operators**—Through their collaborative agreement with the critical infrastructure sector owners and operators will coordinate and communicate directly with the appropriate leadership of critical infrastructure owners/operators. This may include crucial communication with the Multiple Sharing and Analysis Center (ISACs) during a cyber incident.
  - **Federal agencies**—Significant cyber incidents may require nationally-coordinated rapid response actions based on differing authorities and priorities. Numerous organizations may provide essential data and capabilities: DHS, National Security Agency (NSA), DOJ, FBI, Department of State, and other Federal departments and agencies.
  - **State and local government**—Personnel from State, local, tribal, or territorial governments also play a major role in providing relevant information about the sector or entities, through media such as the Multi-State Information Sharing and Analysis Center (MS-ISAC).

- **Einstein 1, 2**—Review the Einstein situational awareness report when studying the cyber incident. Einstein 1, and 2 is a program launched by the National Cyber Security Division (NCSD) in 2004. It is a phased program that adds new functionality to combat cybersecurity exploits to Federal executive agency information technology (IT) enterprises. The first phase, Einstein 1, an intrusion detection system (IDS), was designed to provide situational awareness for civilian agencies by collecting computer network security information such as network flow records, source IP address, port address, communication time, destination IP address, and the port of the computer. Einstein 2, launched in 2008, incorporates network intrusion detection that monitors for malicious activity in the network traffic to and from participating Federal executive agencies.

- **Intelligence analysis**—The purpose of intelligence analysis is to reveal the underlying significance of selected incident information. The Federal government should begin with confirmed and verified incident information based on data source such as the NSA, Central Intelligence Agency, or FBI, and apply expert knowledge to produce plausible data for the initial briefing, final briefing, and reports to the stakeholders.

- **Review updates**—The collection of data and appropriate cyber-incident information from the various internal and public sources, Federal partners, and sector entities listed will be continually revisited and updated as appropriate by the cyber analyst to better assess the risks and consequences from the cyber attack.

2.2   Threat Analysis

Cyber threat analysis is the practice of effectively fusing incident information, knowledge of an organization's network and vulnerabilities—both internal and external, including essential IT and industrial control systems (ICS)—and matching these against other actual cyber attacks and threats that have been observed or reported. The output of this fused analysis is an advanced defensive detection mechanism with a final goal of enhancing the defensive posture of seemingly unaffected or affected asset owners of the critical infrastructure network.

   As part of this process, the cyber analysts are responsible for threat analysis, which will characterize the attack, including scope and scale, from forensic information, and will attempt to ascertain the level of sophistication of the attack and the potential impact to the sector(s). This will include identification of other potentially vulnerable sector(s) and possible detection mechanisms for the attack. The attack's level of severity will be based on the seven layers of the Open Systems Interconnection (OSI) Model described in detail in Appendix D. The steps for threat analysis are as follows:

- **Identify tactics, techniques, and procedures (TTPs)**—The cyber analyst identifies tactics, techniques, and procedures that pertains to cyber attacks. Attacks such as distributed denial of service (DDoS), Internet of Things (IOT) and Ransomware are just few examples that the critical infrastructure sectors operators may experience in the year 2019. For more details on the top 10 types of attacks, see Appendix B.
- **Define scope/scale**—The Federal government should identify the scope and scale of the cyber-physical incident in terms of the interruption of the continuity of daily business activities.
- **Determine the intent and capabilities**—The cyber analyst may seek information from operation centers on the intent, scale, and capabilities of the cyber attack (if available) to determine the type of attack and to understand the impact the disruption will have on the critical infrastructure and the potential defense and detection mechanisms that will be required for mitigation.
- **Identify the sector affected (16 sectors) and any additional sector(s) with the potential to be affected**—The analyst should identify the sector(s) affected by the cyber-physical exploit to understand the upstream and downstream disruption impact on the sector supply chain and other sector(s) dependencies. This should include identification of other sectors that may be similarly vulnerable, so that the appropriate notification can be developed and communicated.

    - **Identify systems affected: cyber or physical**—To understand the impact of a cyber-physical exploit on network infrastructure, it is imperative that information and control systems of the affected infrastructure be identified and understood by the analyst in terms of cyber-physical operational vulnerabilities and consequences.

- – **Determine severity**—Review seven layers of the OSI Reference Model to understand the severity of the cyber attack on the network systems, the cyber analyst should review the seven layers of the OSI Reference Model as a guide for how data are transmitted over the network. The OSI Reference Model is a representation of the critical data pathway that an adversary can exploit. For more information on the OSI Reference Model and figures, see Appendix D.

- • **Identify and review forensics (Infrastructure Protection Packet Capture, Security Information and Event Management [SIEM] Forensic Integration Tool)**—The most common goal of performing network forensics or digital media analysis is to gain a better understanding of an event of interest by finding and analyzing the facts related to that event. Forensics may be needed in many different situations, such as evidence collection for legal proceedings and internal disciplinary actions, and handling of malware incidents and unusual operational problems. See Appendix A for additional cyber incident analysis tools.

Computer and network forensics, or digital media analysis, has evolved to assure proper representation of computer crime evidentiary data in court. National Institute of Standards and Technology (NIST) SP800-86, *Guide to Integrating Forensic Techniques into Incident Response*, describes the forensic or digital media analysis process in terms of collection, examination, analysis, and reporting (Fig. 2.1).

Digital media analysis data using tools and techniques such as the IDS Network Forensic Analysis Tool (NFAT) and SIEM provide incident logs and traces that are useful in determining the type of attack and criminal behavior. Forensics or digital media analysis are most often thought of in the context of criminal investigations and computer security incident handling used to respond to an event by investigating suspect systems, gathering and preserving evidence, reconstructing events, and assessing the current state of an event. Many forensic or digital media analysis tools and techniques can be applied to troubleshooting operational issues, such as finding the virtual and physical location of a host with an incorrect network configuration, resolving a functional problem with an application, and recording and reviewing the current operating system (OS) and application configuration settings for a host. Tools and techniques are available to analyze log entries and recover lost data from

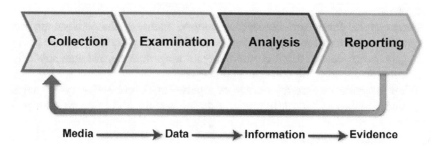

Fig. 2.1 Cyber digital media analysis process (Kent et al. [4])

user and host systems. NIST SP800-86 provides a guide to integrating forensic or digital media analysis techniques into incident response.[4]

- **Review and identify network configuration vulnerabilities (e.g., sensors, firewalls, routers, host, IDS/intrusion protection system [IPS], host anti-virus [AV])**—The cyber analysts should consider reviewing and identifying critical network vulnerabilities during the initial stage of the cyber-incident analysis (if available); these may include misconfigured firewalls, sensors, routers, IDSs and IPSs, and host AV software, as well as risks associated with vendor-supplied software, risks associated with the network, and systems administration errors. If the information is not available in the initial analysis, efforts should be made to seek this information during a latter stage of an analysis, such as digital media analysis, to aid in the development of response and recovery plans. The following are other vulnerabilities that should be considered:

  - *Network vulnerabilities*—Review of network vulnerabilities across information and control systems includes computers, network hardware/systems, OSs, and software applications that may have originated from a vendor system, system administration activities, and/or user activities.
  - *Vendor vulnerabilities*—Vendor-originated vulnerabilities includes software bugs, missing OS patches, vulnerable services, insecure default configurations, and Web applications.
  - *System administration vulnerabilities*—System administration originated vulnerabilities include incorrect or unauthorized system configuration changes and lack of password-protection policies.
  - *User vulnerabilities*—User-originated vulnerabilities include sharing of directories with unauthorized users, failure to run virus scanning software, and malicious activities such as introducing system backdoors.[5]

- **Review network control systems**—Through the Cyber Operation Centers, the analyst leverages capabilities that provide onsite support and mitigation information for protection against and in response to cyber threats; such information may include incident response, forensic analysis, and site assessments. Information and data from ICS-CERT investigative, forensic, or digital media analysis tools can provide situational awareness of evolving threats to an ICS from cyber exploits.
  For example, during the discovery of Stuxnet malware, the network analysis quickly revealed that sophisticated malware of this type potentially has the ability to gain access to, steal detailed proprietary information from, and manipulate the systems that operate mission-critical processes.
- **Filter information through the Cyber Operation Centers**—The cyber analyst should filter the incident and its impact to the sector entity in close cooperation with

---

[4] Kent et al. [4].

[5] How to Scan Backdoors of Your Hacked Word Press by Sufie Banu, June 5, 2018.

the different cyber entities. The operational centers continuously monitor national and international incidents and events that may affect emergency communications.

2.3 Malware Analysis

Malware analysis entails comprehensive review of the regional risk from an infrastructure disruption from cyber threats (see Appendix B). The analysis provides a holistic view of the problem to assist the critical sector(s) in response and recovery. The analysis should consider the following:

- **Collect and analyze the data profile**—Collection and analysis of data should include information needed to analyze and manage critical infrastructure risks. The dataset should include addresses, points-of-contact, asset geo-location, and other information. This data should leverage geographic information system to visually represent the data on a map; it can be sorted by sector, risk, and priority. Output should provide tabular results of the critical infrastructure in question.
- **Perform security and vulnerability assessments**—This step should provide analysts a way to assess the vulnerabilities and consequences of the threat's impact on infrastructure assets. Current cyber threat modeling and analysis tools should provide a comprehensive score based on past and current analyst knowledge and experience. The assessment should allow the analyst to tweak data based on local threat and sector conditions.
- **Separate asset groups for analysis and monitoring**—The malware analysis should allow the analyst to separate and prioritize asset groups based on threat and consequence. These sector assets can be efficiently monitored for manipulation of data based on changing conditions in the field.
- **Document and disseminate results**—The analyst can document the results from the malware analysis and target the results based on a specific audience and according to the different protocols.

2.4 Cyber Incident Threat Information Process

The cyber analyst can seek information from various cyber threat models, which may include threat identification, characterization, and indicator patterns for detection and investigation of specific incidents, in order to determine reactive courses of action. These models are utilized as a structured language for cyber threat intelligence information that capture, characterization, and communication of standardized cyber-threat information; it improves consistency, efficiency, interoperability, and overall situational awareness. A variety of high-level cybersecurity cases rely on such information (Fig. 2.2), which is provided by cyber analysis, which works in conjunction with Common Indicator Threat Information. See Appendix C for more

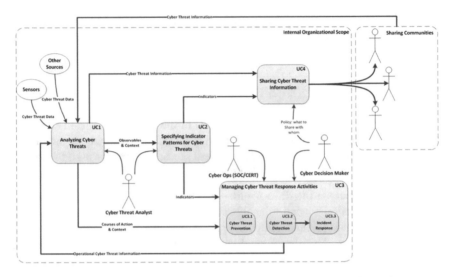

Fig. 2.2 Notional core use cases targeted by cyber indicators™ (https://stix.mitre.org)

information on cyber analysis processes. Analytic components to consider in this process include:

- **Analyzing cyber threats**—A cyber threat analyst reviews structured and unstructured information regarding cyber threat activity from a variety of manual or automated input sources. The analyst seeks to understand the nature of relevant threats, identify them, and fully characterize them such that all of the relevant knowledge of the threat can be fully expressed and evolved over time. This relevant knowledge includes threat-related actions, behaviors, capabilities, intents, attributed actors, and other information.
- **Specifying indicator patterns for cyber threat**—A cyber threat analyst specifies measurable patterns that represent the observable characteristics of specific cyber threats, as well as their threat context and relevant metadata for interpreting, handling, and applying the pattern and its matching results.

For example, in the case of a confirmed phishing attack, an analyst may harvest the relevant set of observables (e.g., to or from addresses, actual source, subject, embedded uniform resource locators [URLs], type of attachments, specific attachment) from the performed analysis of the phishing email, identify the relevant TTPs exhibited in the phishing attack, perform a kill-chain correlation of the attack, assign appropriate confidence for the indicator, determine appropriate handling guidance, generate any relevant automated rule patterns for the indicator (e.g., Snort, OVAL), assign any suggested courses of action, and package it as a coherent record for sharing.

- **Managing cyber threat response activities**—Cyber decision-makers and cyber operations personnel work together to prevent or detect cyber threat activity and to investigate and respond to any detected incidences of such activity. Preventative courses of action may be remedial in nature to mitigate vulnerabilities, weaknesses, or misconfigurations that may be targets of exploit. After detection and investigation of specific incidents, reactive courses of action may be pursued. For example, in the case of a confirmed phishing attack with defined indicators, decision-makers and personnel work together to fully understand the effects of the phishing attack within the environment, including malware installed or malware executed; to assess the cost and efficacy of potential courses of action; and to implement appropriate preventative or detective courses of action.

- **Sharing cyber threat information**—Cyber decision-makers establish policy for what sorts of cyber threat information will be shared with which other parties and how such information should be handled based on agreed-to frameworks of trust to maintain appropriate levels of consistency, context, and control. This policy is then implemented to share the appropriate cyber threat indicators and other cyber threat information. For example, in the case of a confirmed phishing attack with defined indicators, the policies predefined by cyber decision-makers could allow the relevant indicators to be automatically or manually shared with trusted partners or communities so they could take advantage of the knowledge gained by the sharing organization.[6]

References

1. Federal: Cybersecurity and Infrastucture Agency (CISA) (2019) (undated a). us-cert.gov/NCC ICReports/year-in-Review_Final_5500BC.pdf
2. Federal: Bottom-Up Review Report (2019). www.ics-cert.us-cert.gov/sites/default/files/Annual
3. Federal (U.S. Department of Homeland Security): Federal daily open source infrastructure report. Available at http://www.dhs.gov/dhs-daily-open-source-infrastructure-report (2013). Accessed 25 Mar 2013
4. Kent, K., Chevalier, S., Grance, T., Dang, H.: Special publication SP800-86, guide to integrating forensic techniques into incident response. NIST (2006, August). Available at http://csrc.nist.gov/publications/nistpubs/800-86/SP800-86.pdf. Accessed 25 Mar 2013

[6]The MITRE Corporation, 2018, mitre.org.>capabilities>cybersecurity>cyberthreatintelligence.

Chapter 3
Vulnerability/Consequence Analysis

3.1 Information Sharing

Secure, functioning, and resilient critical networks requires the efficient exchange of information, including intelligence, between all levels of government and critical infrastructure sector owners and operators. This must facilitate the timely exchange of threat and vulnerability information as well as information that allows for the development of a situational awareness capability during incidents.[1]

To conduct vulnerability/consequence analysis of any particular sector or potential exploit, the analyst will collect relevant data from cyber operational centers on credible cyber threats and combine it with cyber-physical system information from owner/operators. Collection of data includes understanding potential cyber exploit details, potential impacts on the entity based on their information, and relevant mitigation steps. The analyst will leverage partnerships with the government, industry, and international partners to determine risk to infrastructure owners and operators from credible cyber threats. These analyses will provide the basis for pre-incident preparedness activities and post-incident response and recovery.

3.2 Vulnerability/Consequence Analysis

To conduct a vulnerability/consequence analysis, a method for assessing and rating the risk (low, medium, and high) of possible cyber vulnerability for a specific critical infrastructure facility is needed. The risk is a function of the likelihood (probability) that a defined threat agent (adversary) who has the intent and capabilities can exploit specific cyber-physical vulnerability and create an impact (consequence). The risk induced by any given cyber vulnerability is influenced by a number of related indicators, including the following:

---

[1]Whitehouse.gov [1].

© Springer Nature Switzerland AG 2021
Fiedelholtz, *The Cyber Security Network Guide*, Studies in Systems,
Decision and Control 274, https://doi.org/10.1007/978-3-030-61591-8_3

- **Policy and procedures**—Vulnerabilities are often introduced into cyber network because of incomplete, inappropriate, or nonexistent security documentation, including policy and implementation guides, proper change management, and procedures.
- **Computer architecture and conditions**—Vulnerabilities in networks can occur due to flaws, misconfigurations, or poor maintenance (change management process) of their platforms, including hardware, OS, and updated software/applications.
- **Network architecture**—Vulnerabilities in networks may occur from flaws, misconfigurations, or poor administration of networks and their connections with other networks.
- **Installed countermeasures**—Care must be taken before using tools such as an IDS or IPS to identify and protect network applications from malware. Without assessment and proper software updates, these countermeasures may have an adverse impact on the production of hardware and software configuration of the network.
- **Technical difficulty of the attack**—Attacks on the ICS network (reliant on third-party contractor support due to technical complexities of the project) by sophisticated malware such as Stuxnet, which is designed to be stealthy, are examples of the technical difficulty of an attack.
- **Probability of detection**—Probability of detection is important because security is built on a layered defense. If there is a 3% chance that an exploit can make it into a network and a 4% chance an exploited system will go undetected, then there is only a 0.12% chance that someone will manage both feats at the same time. It is impossible to reduce the probability all the way to 0%, but it is possible to make the probability very small. Another example of probability of detection is the amount of time the adversary can remain in contact with the target ICS or network without detection.
- **Consequences of the incident**—A cyber breach in some critical infrastructures can have significant physical impacts (personal injury and loss of life), as well as economic (greater economic loss on the facility and or organization), and social (loss of national or public confidence in an organization) impacts.
- **Cost of the incident**—Undesirable incidents of any sort detract from the value of an organization, but safety and security incidents can have longer-term negative impacts than other types of incidents on all stakeholders, including employees, shareholders, customers, and the communities in which an organization operates.[2] For example, the *Hudson Valley Times* reported on February 28, 2013, the Central Hudson Gas & Electric Company on the cyber attack to their financial system. It was suspected that the cybersecurity breach may have compromised financial data of more than 100,000 of the utility's customers. The company offered potentially impacted customers a year of free credit monitoring and advised them to keep

---

[2]Stouffer et al. [2].

an eye out for suspicious activity on their bank accounts to avoid loss of public confidence.[3]

3.2.1 Collect Cyber Data

- Collection of validated cyber incident and forensic data involves the search for, recognition of, collection of, and documentation of all the electronic evidence collected by NCCIC automated tools and the sector participant.
- **Network and host information**—Cyber data collection involves review of network and host information collected from application servers, system log files, firewall log records, IDS, AV, malware detection automated tools, ISP log files, and interviews of people and departments involved in the incident. For example, the Network Mapper (Nmap) open-source tool for network exploration and security auditing can provide information about hosts available on the network, types of services (application names and version), OS versions, types of packet filters/firewalls that are in use, and other characteristics.

3.2.2 Physical Analysis of Cyber Controlled/Reliant Systems

- To conduct the vulnerability and consequences analysis, it is imperative for the analyst to understand the physical and cyber attributes of the cyber-controlled reliant systems. This section provides an overview of supervisory control and data acquisition (SCADA), distributed control system (DCS), and programmable logic controller (PLC) systems, including typical architectures and components. To facilitate understanding of these systems, a network diagram is presented to depict the network connections and components typically found in each system. However, it is important to remember that actual implementations of ICSs may be hybrids that blur the line between DCSs and SCADA systems by incorporating attributes of both.
- **ICS/DCS**—ICS is a general term that encompasses several types of control systems, including SCADA systems, DCSs, and process control systems configurations such as the skid-mounted PLCs often found in the industrial sector and critical infrastructures. ICSs are typically used in industries such as electrical, water and wastewater, oil and natural gas, chemical, transportation, pharmaceutical, pulp and paper, food and beverage, and discrete manufacturing (e.g., automotive, aerospace, and durable goods). These control systems are critical to the operation of U.S. critical infrastructure assets that are often highly interconnected and mutually dependent.
  SCADA systems are highly distributed systems used to control geographically dispersed assets, often scattered over thousands of square kilometers, where

---

[3] *Hudson Valley Times* [3].

centralized data acquisition and control are critical to system operation. They are used in distribution systems such as water distribution and wastewater collection systems, oil and natural gas pipelines, electrical power grids, and railway transportation systems. A SCADA control center performs centralized monitoring and control for field sites over long-distance communications networks, including monitoring alarms and processing status data. Based on information received from remote stations, automated or operator-driven supervisory commands can be pushed to remote station control devices, which are often referred to as field devices.

Field devices control local operations such as opening and closing valves and breakers, collecting data from sensor systems, and monitoring the local environment for alarm conditions.

DCSs are used to control industrial processes such as electric power generation, oil refineries, water and wastewater treatment, and chemical, food, and automotive production. DCSs are integrated as a control architecture containing a supervisory level of control overseeing multiple integrated subsystems that are responsible for controlling the details of a localized process. Product and process control are usually achieved by deploying wired or wireless feedback or feed forward control loops whereby key product and/or process conditions are automatically maintained around a desired set point. To accomplish the desired product and/or process tolerance around a specified set point, specific PLCs are employed in the field and proportional, integral, and/or derivative settings on the PLC are tuned to provide the desired tolerance as well as the rate of self-correction during process upsets.

DCSs are used extensively in process-based industries in conjunction with SCADA systems to remotely operate multiple processing facilities. An example of the wireless SCADA for a water treatment facility is shown in Fig. 3.1.

PLCs are computer-based solid-state devices that control industrial equipment and processes. While PLCs are control system components used throughout SCADA systems and DCSs, they are often the primary components in smaller control system configurations used to provide operational control of discrete processes such as automobile assembly lines and power plant soot blower controls. PLCs are used extensively in almost all industrial processes.[4]

The following is a list of the generic ICS and communication network architecture components (Fig. 3.2) that are vulnerable to cyber exploits:

- **Control server**—The control server hosts the DCS or PLC supervisory control software that communicates with lower-level control devices. The control server accesses subordinate control modules over the ICS network.
- **SCADA server or master terminal unit (MTU)**—The SCADA server is the device that acts as the master in a SCADA system. Remote terminal units (RTUs) and PLC devices (as described below) located at remote field sites usually act as slaves.

---

[4]Stouffer et al. [4].

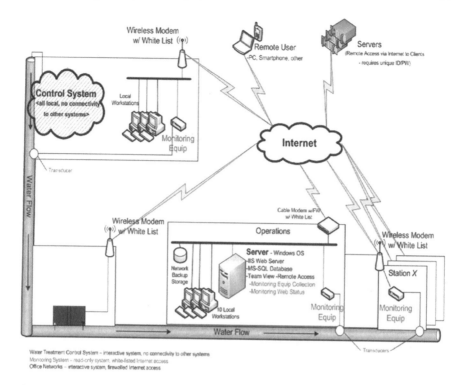

Fig. 3.1 Notional critical infrastructure network layout facility (NIST 800-82, Revision guide to control systems)

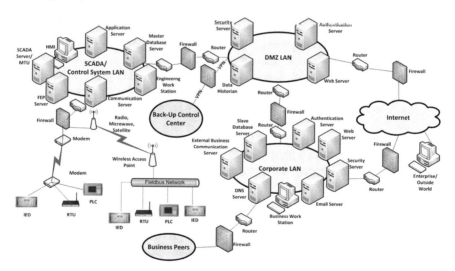

Fig. 3.2 Notional SCADA/Control System Network Architecture (https://inductiveautomation.com)

- **RTU**—The RTU, also called a remote telemetry unit, is a special-purpose data acquisition and control unit designed to support SCADA remote stations. RTUs are field devices often equipped with wireless radio interfaces to support remote situations where wire-based communications are unavailable. Sometimes PLCs are implemented as field devices to serve as RTUs; in this case, the PLC is often referred to as an RTU.
- **PLC**—The PLC is a small industrial computer originally designed to perform the logic functions executed by electrical hardware (relays, switches, and mechanical timer/counters). PLCs have evolved into controllers with the capability of controlling complex processes and are used substantially in SCADA systems and DCS. Other controllers used at the field level include process controllers and RTUs; they provide the same control as PLCs but are designed for specific control applications. In SCADA environments, PLCs are often used as field devices because they are more economical, versatile, flexible, and configurable than special-purpose RTUs.
- **Intelligent electronic device (IED)**—An IED is a "smart" sensor/actuator containing the intelligence required to acquire data, communicate to other devices, and perform local processing and control. An IED could combine an analog input sensor, analog output, low-level control capabilities, a communication system, and program memory in one device. The use of IEDs in SCADA systems and DCSs allows for automatic control at the local level.
- **Human-machine interface (HMI)**—The HMI incorporates software and hardware that allow human operators to monitor the state of a process under control, modify control settings to change the control objective, and manually override automatic control operations in the event of an emergency. The HMI also allows a control engineer or operator to configure set points or control algorithms and parameters in the controller. The HMI also displays process status information, historical information, reports, and other information to operators, administrators, managers, business partners, and other authorized users. Location, platform, and interface may vary widely. For example, an HMI could be a dedicated platform in the control center, a laptop on a wireless local-area network (WLAN), or a browser on any system connected to the Internet.
- **Data historian**—The data historian is a centralized database for logging all process information within an ICS. Information stored in this database can be accessed to support various analyses, from statistical process control to enterprise-level planning.
- **Input/Output (I/O) server**—The I/O server is a control component responsible for collecting, buffering, and providing access to process information from control sub-components such as PLCs, RTUs, and IEDs. An I/O server can reside on the control server or on a separate computer platform. I/O servers are also used for interfacing third-party control components, such as an HMI and a control server.
- Network topologies and cross-domain solutions across different ICS implementations vary with modern systems that use Internet-based IT and enterprise

integration strategies. Control networks have merged with corporate networks to allow control engineers to monitor and control systems from outside of the control system network. The connection may also allow enterprise-level decisionmakers to obtain access to process data. The following is a list of additional major components and considerations of an ICS network, regardless of the network topologies in use:

- **Network cross-domain solutions**—The network cross-domain solutions are crucial in the integration of a SCADA system or ICS to ensure that the software and the hardware configuration interface are working in tempo so the netflow real-time traffic can run effectively.
- **Fieldbus network**—The fieldbus network links sensors and other devices to a PLC or other controller. Use of fieldbus technologies eliminates the need for point-to-point wiring between the controller and each device. The devices communicate with the fieldbus controller using a variety of protocols. Messages sent between the sensors and the controller uniquely identify each of the sensors.
- **Control network**—The control network connects the supervisory-level control to lower-level control modules.
- **Communications router**—A router is a communications device that transfers messages between two networks. Common uses for routers include connecting a local area network (LAN) to a wide area network (WAN), and connecting MTUs and RTUs to a long-distance network for SCADA communication.
- **Firewall**—A firewall protects devices on a network by monitoring and controlling communication packets using predefined filtering policies. Firewalls are also useful in managing ICS network segregation strategies.
- **Demilitarized zone (DMZ)**—DMZ is a firewall configuration for securing a LAN. In a DMZ configuration, most computers on the LAN operate behind a firewall connected to another network such as a business network, the Internet, or another type of network. The computers in the DMZ intercept data traffic and broker requests for the rest of the LAN, adding an extra layer of protection for the computers behind the firewall.[5]
- **Modem**—A modem is a device used to convert between serial digital data and a signal suitable for transmission over a telephone line to allow devices to communicate. Modems are often used in SCADA systems to enable long-distance serial communications between MTUs and remote field devices. They are also used in SCADA systems, DCSs, and PLCs for gaining remote access for operational and maintenance functions such as entering commands or modifying parameters, and diagnostic purposes.
- **Remote access points**—Remote access points are distinct devices, areas, and locations of a control network for remotely configuring control systems and accessing process data. Examples include using a personal digital assistant

---

[5]https://www.lifewire.com, June 2019.

(PDA) to access data over a LAN through a wireless access point, and using a laptop and modem connection to remotely access an ICS.[6]

Additional systems include the following:

- **Safety instrumented system (SIS)**—A SIS plays a vital role in providing the protective layer functionality in many critical infrastructure industrial process and automation systems. Automatic SIS are some of the mitigation measures the industry uses to protect against or mitigate harm/damage to personnel, process plant, and the environment. A SIS comprises sensors, logic solvers, and actuators to take a process to a safe state when normal predetermined set points are exceeded, or safe operating conditions are violated. In general, any protection system (including a SIS) is kept functionally separate from the process control systems so that it can operate independently of the state of the control systems. In essence, protection systems such as SISs should be capable of functioning to protect the processes under their control when the control system fails or is in fault mode.[7] Today, due to the complex process and automation systems, separation of the SIS from process control system is not possible and therefore could be exploited using malware, or when the process control system is compromised by a perpetrator.
- **Emergency shutdown system**—An emergency shutdown (ESD) system is an automated safety system layer of protection designed to mitigate and prevent plant processes, personnel, and the environment from a hazardous situation. The ESD is designed to shut down a process in a safe and orderly fashion during an emergency. An emergency shutdown system for a process control system includes an ESD valve and an associated valve actuator. An ESD controller provides output signals to the ESD valve in the event of a failure in the process control system. A solenoid valve responds to the ESD controller to vent the actuator to a fail state. A digital valve controller (DVC) test strokes the ESD valve. An impedance booster device allows the direct current powering of the solenoid valve and the DVC over a two-wire line while still permitting digital communication over the same two-wire line.[8]
- The ESD system can be activated manually or automatically from a third-party system such as a fire and gas system or process sensors. Once activated, the ESD system initiates customer-defined actions, which may consist of either partial or total plant shutdown procedures. It is not uncommon for the industry to transmit shutdown signal, defined system, and discrepancy alarms data using a third-party DCS network and backup communication systems architecture, control, and monitoring devices. The following are components that may be part of the ESD system:

---

[6]Stouffer et al. [4].

[7]https://Automation Forum Co, May 24, 2018.

[8]Esoteric Automation & Control Technologies, undated, "Emergency Shut Down (ESD) System, available at https://pdfs.semanticsscholar.org.

– **Cyber-physical controls**—A cyber-physical control system integrates computing and communication capabilities with the monitoring and/or control of physical industrial and business processes. Embedded computers, control systems, and communication networks monitor and control physical processes such as generators and pumps. This usually involves feedback loops in which physical processes affect computations, and vice versa, providing information such as alerts and alarms to the control center operator.

> **Generators/pumps**—Generators and operations of pumps depend on control systems such as DCSs to provide reliable means for control, operational efficiency, and generation process optimization. The DCS architecture includes LAN and corporate WAN, HMI, hardware, and software to allow an operator to control physical equipment such as generators and pumps through cyber means.
>
> **Alerts/alarms**—Examples of cyber-physical controls that are used to support control system security on the physical process include alarms/alerts on user accounts, AV, security appliance devices, routers and firewalls, IDS, audit trail systems, and others.

- **Communications**—Communication protocols and communication mediums control system environments use for field device control with central control center communication are typically different from the generic IT environment. They may be proprietary, but control systems are increasingly migrating toward the use of IP for interoperability and application of open vendor systems. In general, DCS communications use LAN technologies, while SCADA communications use long-distance communication such as microwave, public telephone network, and satellite systems. Communications to field devices often use industry standard protocols such as Distributed Network Protocol 3.0 and Modbus. These protocols were originally developed to run over serial connections, but were layered on top of transmission control protocol (TCP)/IP for the convenience and efficiency of LAN/WAN communications. Cyber exploits such as DDoS can potentially interrupt communication between devices and control center and make it difficult for the sector operator to control and monitor the processes.
- **Control systems**—Adversaries who can monitor the control system network activity can use a protocol analyzer or other utilities to decode the data transferred by protocols such as telnet, File Transfer Protocol (FTP), and Network File System (NFS). The use of such protocols also makes it easier for adversaries to perform attacks against the control system and manipulate control system network activity. Many control system protocols have no authentication at any level. Without authentication, there is the potential to replay, modify, or spoof data or to spoof devices such as sensors and user identities.[9]

---

[9]Stouffer et al. [4].

3.3   Dependency/Interdependency Analysis

The term *control system* encompasses several types of systems, including SCADA, process control, and other automated systems that are found in the industrial sectors and critical infrastructure. These systems operate physical processes that produce the goods and services the Nation relies upon, such as energy, drinking water, emergency services, transportation, postal and shipping, and public health. Infrastructure dependency/interdependency can be described in terms of four general categories:

- Physical (e.g., the material output of one infrastructure is used by another),
- Cyber (e.g., infrastructures utilize electronic information and control systems),
- Geographic (e.g., infrastructures are collocated in a common corridor), and
- Logical (e.g., infrastructures are linked through financial markets).

Control systems security is particularly important because of the inherent interconnectedness of the critical infrastructure sectors and their dependence on one another. As such, assessing risk and effectively securing ICSs are vital to maintaining the Nation's strategic interests, public safety, and economic wellbeing. A successful cyber attack on a control system could result in physical damage, loss of life, and cascading effects that could disrupt services.[10]

3.3.1   Identify Internal Impacts

The U.S. critical infrastructure is often referred to as a "system of systems" because of the interdependencies that exist between its various industrial sectors as well as interconnections between business partners. Critical infrastructures are highly interconnected and mutually dependent in complex ways, both physically and through a host of information and communications technologies. An incident in one infrastructure can directly and indirectly affect other infrastructures through cascading and escalating failures. Definitions for dependencies and interdependencies are as follows:

- **Dependencies**—Critical infrastructures are highly interconnected and mutually dependent in complex ways, both physically and through a host of information and communications technologies. An incident in one part of the infrastructure can directly and indirectly affect other infrastructures through cascading and escalating failures. Certain physical systems are crucial to the functioning of the delivery of the goods and services of the systems.
- **Interdependencies**—Electric power is often considered one of the most prevalent sources of disruptions for interdependent critical infrastructures. Impacts or

---

[10]https://iecetech.org/index.php/Technology-Focus2019-02/cyberattacks-targeting-criticalinfrast ructure.

failures affecting the interdependent infrastructures can be described in terms of three general categories:

- **Cascading failure**—A disruption in one infrastructure causes a disruption in a second infrastructure.
- **Escalating failure**—A disruption in one infrastructure exacerbates an independent disruption of a second infrastructure (e.g., the time for recovery or restoration of an infrastructure increases because another infrastructure is not available).
- **Common cause failure**—A disruption of two or more infrastructures at the same time is the result of a common cause (e.g., natural disaster).

Figure 3.3 shows an illustration of cascading and escalating failures for representative components of the electric power and natural gas infrastructures. As depicted, a cascading failure is initiated by a disruption of the microwave communications network that is used for the SCADA system. The lack of monitoring and control capabilities causes a large generating unit to be taken offline, an event that, in turn, causes a loss of power at a distribution substation. This loss then leads to blackouts for the area served by the substation. The outages affect traffic signals; this problem

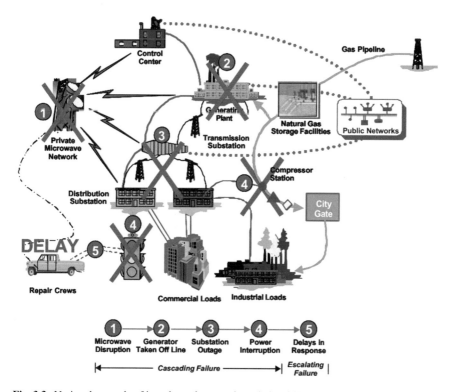

Fig. 3.3 Notional example of interdependency and escalating failures (McCauley [5])

increases travel times and causes delays in repair and restoration activities (escalating failure). In addition, the electric power loss could cause a major imbalance, triggering a cascading failure across the power grid. This could result in large area blackouts that could potentially affect oil and natural gas production, refinery operations, water treatment systems, wastewater collection systems, and pipeline transport systems that rely on the grid for electric power.[11]

- **Resiliency measures**—Infrastructure resilience is the ability to reduce the magnitude and/or duration of disruptive events. The effectiveness of a resilient infrastructure or enterprise depends upon its ability to anticipate, absorb, adapt to, and/or rapidly recover from a potentially disruptive event. The National Infrastructure Advisory Council (NIAC) has the following resiliency measures for critical infrastructure:

  - **Absorptive capacity** is the ability of the system to endure a disruption without significant deviation from normal operating performance. For example, fireproofing foam increases the capacity of a building system to absorb the shock of a fire.
  - **Adaptive capacity** is the ability of the system to adapt to a shock affecting normal operating conditions. For example, the extra transformers U.S. electric power companies keep on store and share increases the ability of the grid to adapt quickly to regional power losses.
  - **Recoverability** is the ability of the system to recover quickly—and at low cost—from potentially disruptive events.[12]

3.3.2 Identify External Impacts

The cyber analyst should identify external impacts from a cyber-physical perspective. All critical infrastructure assets depend on, and are needed for, one or more other critical infrastructures. For an example, the chemical sector depends on, and is needed for, a wide range of other sectors, including communications, critical manufacturing, emergency services, energy, food and agriculture, healthcare and public health, IT, transportation systems, and water and wastewater systems.[13] The highly simplified example (Fig. 3.4) shows how to determine external impacts from a cyber incident to understand and analyze cascading or escalating failures. The following areas should be considered when identifying external impacts:

- **Dependencies**—The 16 critical infrastructures are now increasingly dependent on automation technology, as depicted in Fig. 3.4. To evaluate the cyber-physical impacts from a cyber incident, it is vital that the analyst understand the relationship and implications of dependencies and interdependencies among the

---

[11] Stouffer et al. [4].

[12] He and Chia [6].

[13] Chemical Sector [7].

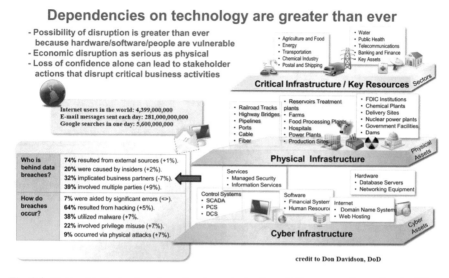

Dependencies on technology are greater than ever

- Possibility of disruption is greater than ever because hardware/software/people are vulnerable
- Economic disruption as serious as physical
- Loss of confidence alone can lead to stakeholder actions that disrupt critical business activities

Internet users in the world: 4,399,000,000
E-mail messages sent each day: 281,000,000,000
Google searches in one day: 5,600,000,000

Critical Infrastructure / Key Resources Sectors

- Agriculture and Food
- Energy
- Transportation
- Chemical Industry
- Postal and Shipping
- Water
- Public Health
- Telecommunications
- Banking and Finance
- Key Assets

Physical Infrastructure Physical Assets

- Railroad Tracks
- Highway Bridges
- Pipelines
- Ports
- Cable
- Fiber
- Reservoirs Treatment plants
- Farms
- Food Processing Plants
- Hospitals
- Power Plants
- Production Sites
- FDIC Institutions
- Chemical Plants
- Delivery Sites
- Nuclear power plants
- Government Facilities
- Dams

Who is behind data breaches?	**74%** resulted from external sources (+1%).
	20% were caused by insiders (+2%).
	32% implicated business partners (-7%).
	39% involved multiple parties (+9%).
How do breaches occur?	**7%** were aided by significant errors (<>).
	64% resulted from hacking (+5%).
	38% utilized malware (+7%.
	22% involved privilege misuse (+7%).
	9% occurred via physical attacks (+7%).

Cyber Infrastructure Cyber Assets

Services
- Managed Security
- Information Services

Control Systems
- SCADA
- PCS
- DCS

Software
- Financial System
- Human Resource

Hardware
- Database Servers
- Networking Equipment

Internet
- Domain Name System
- Web Hosting

credit to Don Davidson, DoD

Fig. 3.4 Notional infrastructure dependencies on technology (Infrastructure Dependencies [8])

critical infrastructures for thorough analysis and reporting of vulnerability and consequences.

- **Interdependencies**—Traditionally, interdependencies have been predominantly physical and geographic in nature. However, the proliferation of IT, along with increased use of automated monitoring and control systems and increased reliance on the open marketplace to purchase and sell infrastructure commodities and services, has increased the prevalence and importance of cyber and logical interdependencies. Figure 3.5 shows examples of infrastructure dependencies for the electric power infrastructure. Each of the linkages shown has important, and potentially different, spatial, temporal, and system characteristics.
- Extending this dependency notion to multiple infrastructures, as shown in Fig. 3.6 depicts infrastructure interdependencies from a "system of systems" perspective. The complexity of multiple infrastructure linkages, and the implications of multiple contingency events that may affect the infrastructures, are apparent even in this highly simplified representation.
- **Resilience measures**—The state of operation of an infrastructure in terms can range from normal operation to various levels of stress, disruption, or repair and restoration. These aspects must be considered in examining interdependencies and resilience measures. For example, it is necessary to understand both backup systems and other mitigation mechanisms that reduce interdependence problems, as well as the change in interdependencies as a function of outage duration and frequency.

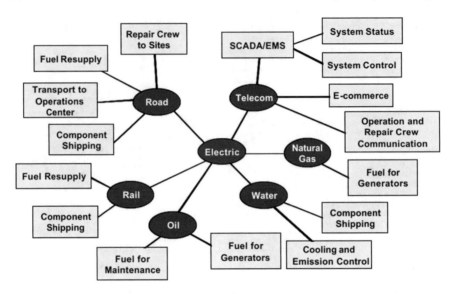

Fig. 3.5 Notional infrastructure interdependencies for electric power (Petit [9])

3.4 Analysis Reporting

Analysis reporting should consist of the following:

- **Conduct vulnerability and consequence analysis**—Systems that control the critical infrastructure in most cases operate constantly and the impact of downtime from a cyber-exploit of the control systems that potentially could endanger public health and safety can range from inconvenient to catastrophic. A vulnerability and consequence assessment of IT and control systems for the sector(s) would identify and report noted vulnerabilities (security weaknesses of the target sector system) and consequences from all the data collected within this process.[14] In collaboration with partners, the analyst should review all aspects of the exploit threat for known security weaknesses. Many security tools and techniques used by cyber operational centers help identify and validate vulnerabilities. Analysis reporting to appropriate leadership and stakeholders should be validated with and sector operators for accuracy and completeness.
- **Conduct critical infrastructure and key resources analysis**—Critical infrastructure experiencing the cyber incident in question will be analyzed, including those assets, systems, networks, and functions (physical or virtual) vital to the Nation, whose incapacitation or destruction would have a debilitating impact on security, national economic security, public health or safety, or any combination of these. Key resources are publicly or privately controlled resources that are essential to minimal operation of the economy and the government. Critical

---

[14]Industrial Cyber Security-campaign.abb.com/industrial/cybersecurity, 2019.

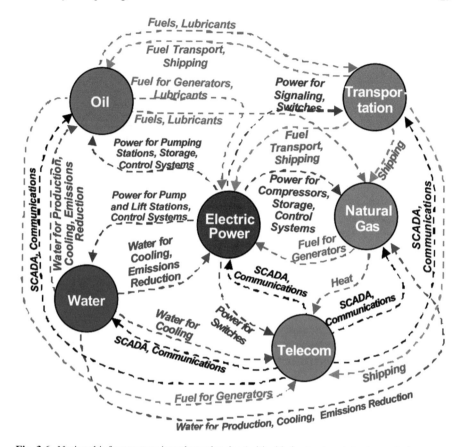

Fig. 3.6 Notional infrastructure interdependencies (critical infrastructure interdepencies)

infrastructure is organized into 16 sectors under the Executive Order "Improving Critical Infrastructure Cybersecurity" that together provide essential functions and services supporting various aspects of the U.S. government, economy, and society.

- **Perform vulnerability analysis**—Performing a vulnerability analysis is a collaborative effort between the public- and private-sector personnel impacted from the cyber incident. The analysis process should utilize all relevant public/private data on the cyber incident to perform an assessment of the infrastructure affected, using modeling and simulation tools to support the vulnerability analysis.
- **Conduct pre-assessment**—The pre-vulnerability assessment should establish collection of all incident-related datasets, assessment of damage from the cyber incident to the relevant entity or infrastructure, vulnerability of the operation, communication protocols with key industry and government partners, situation analysis, areas of concerns/issues, and initial reporting.

- **Conduct assessment**—The vulnerability assessment should utilize the NIPP risk management framework to analyze the implications of the cyber incident pertaining to the affected entity or the critical infrastructure on a local, State, regional, or national basis. The assessment should include threat, vulnerability, and consequence interdependencies; cascading or secondary effects on the entities critical systems or infrastructure; and impact of the incident on the inside operation and outside the affected area. The purpose of this step is to develop a list of system vulnerabilities (flaws or weaknesses) that could be exploited further by the cyber exploits and to determine comprehensive consequences during further analysis activities.
- **Conduct post-assessment**—Similar to the full vulnerability assessment in the previous step, the post-assessment should take into account the observed impacts on the operation of the entity or critical infrastructure against the pre- and full assessment results, identifying areas not addressed, requirement of unavailable datasets, lessons learned, and so on. The lessons-learned process should strive to identify system vulnerabilities and how to prevent future attacks on the system, prevent or limit disruption, and establish early visibility and clear communication of such attacks through enhanced awareness, security monitoring, and information sharing.
- **Perform consequence analysis**—Performance of a consequence analysis as a result of a cyber-physical incident can be described in terms of loss or degradation of any, or a combination of any, of the following three security goals for the entity or critical infrastructure: integrity, availability, and confidentiality. The following list provides a brief description of each security goal and the consequence (or impact) of its not being met:

  - **Integrity**—System and data integrity refers to the requirement that information be protected from improper modification. Integrity is lost if unauthorized changes are made to the data or IT/control system by either intentional or accidental acts. If the loss of system or data integrity is not corrected, continued use of the contaminated system or corrupted data could result in inaccuracy, fraud, or erroneous operations decisions. In addition, violation of integrity may be the first step in a successful attack against system availability or confidentiality. For all these reasons, loss of integrity reduces the assurance of an IT/control system.
  - **Availability**—The organization's mission may be affected if a mission-critical IT/control system is unavailable to its end users. Loss of system functionality and operational effectiveness, for example, may result in loss of productive time, thus impeding the end users' performance of their functions in supporting the organization's mission.
  - **Confidentiality**—System and data confidentiality refers to the protection of information from unauthorized disclosure. The impact of unauthorized disclosure of confidential information can range from the jeopardization of

national security to the disclosure of Privacy Act data. Unauthorized, unanticipated, or unintentional disclosure could result in loss of public confidence, embarrassment, or legal action against the organization.

Determine and implement a method to measure possible consequences of the identified incident.

Prioritize each asset threat against possible local, State, regional, and national consequences.

Translate qualitative/quantitative ratings to numeric values where feasible.

- **Report vulnerability/consequence analysis**—Once the risk assessment has been completed (threat sources and vulnerabilities identified, and consequences assessed), the results should be documented based on the target stakeholders. The report should be documented in a manner that assists senior management and the mission asset/critical infrastructure owners in making decisions on policy, procedural, budget, and system operational and management changes. The report should be presented in a systematic and analytical way so that senior management will understand the incident threat, vulnerabilities, and consequences in order to allocate resources and provide appropriate guidance to the asset owner of the critical infrastructure.[15]

References

1. Whitehouse.gov: Presidential policy directive—critical infrastructure security and resilience. Available at http://www.whitehouse.gov/the-press-office/2013/02/12/presidential-policy-directive-critical-infrastructure-security-and-resil. Accessed 25 Mar 2013
2. Stouffer, K., Falco, J., Scarfone, K.: Special publication SP800-82: guide to industrial control systems (ICS) security. NIST (2011). Available at http://csrc.nist.gov/publications/nistpubs/800-82/SP800-82-final.pdf. Accessed 25 Mar 2013
3. *Hudson Valley Times*: Hack attack. February 28. Available at http://www.ulsterpublishing.com/view/full_story/21844700/article–Hack-attack-? (2013). Accessed 25 Mar 2013
4. Stouffer, K., Falco, J., Scarfone, K.: Special publication SP800-82: guide to industrial control systems (ICS) security. NIST (2015). Available at http://csrc.nist.gov/publications/nistpubs/800-82/SP800-82-final.pdf. Accessed 25 Mar 2013
5. McCauley: The danger of critical infrastructure interdependency (2019)
6. He, X., Chia, E.J.: Modeling the post-disaster recovery of interdependent critical infrastructure network (2019)
7. Chemical Sector: Available at http://www.dhs.gov/chemical-sector. Accessed 25 Mar 2018
8. Infrastructure Dependencies: The big rocks panel—private sector perspective. Svailable at http://www.dtic.mil/ndia/2011DIBCIP/Copeland.pdf. Accessed 25 Mar 2013
9. Petit, F.: Critical Infrastructure Interdependencies OSTI.gov (2016). https://www.osti.gov.>servlets>puri
10. Johnson, C.: NIST special publications, 800-15, guide to cyber threat infrastructure sharing (2016)

---

[15]Johnson [10].

Chapter 4
Incident Response and Recovery

The cyber analyst in the cyber-physical response and recovery will provide necessary support in the analysis and development of response and recovery action plans to protect the sector entity or infrastructure property, business needs, and the environment after an incident has occurred and then continue this effort in stabilizing the incident. Response and recovery efforts are focused on ensuring that the sector entity or infrastructure is able to effectively respond to any threats with an emphasis on economy, environment, and safety.

The ensuing recovery process includes those capabilities necessary to assist the sector entity or infrastructure asset, as well as communities affected by an incident, in recovering effectively and prioritizing action plans and support. It is focused on timely restoration, strengthening, and revitalization of the infrastructure. Successful recovery requires informed and coordinated leadership, collaboration between the sector partners during all phases of the recovery process.

4.1 Information Sharing

4.1.1 Cyber Incident Response

The cyber analyst working with the operational centers develops responses commensurate with levels of response (levels 1, 2, and 3 as defined under Cyber Operational CONOPS) as appropriate for the incident under investigation. Information sharing includes keeping the leadership of International, Federal agencies, and sector partners informed.

© Springer Nature Switzerland AG 2021
Fiedelholtz, *The Cyber Security Network Guide*, Studies in Systems,
Decision and Control 274, https://doi.org/10.1007/978-3-030-61591-8_4

4.1.2 Notify Authority of Cyber Operation Center

In collaboration the cyber analyst provides information on the incident and the response and recovery action items initially drafted to the appropriate leadership.

4.1.3 Review and Provide Feedback from the Cyber Operational Center

The cyber analyst reviews feedback from the operational center on the response and recovery plan and action items. During this step, the cyber analyst shares information and collaborates with other analysts at the operation centers.

4.1.4 Coordinate for Cyber-Physical Analysis

To conduct an incident analysis, the cyber analyst will collect relevant data from operation centers on the incident and collaborate and share information on the sector infrastructure and specifics of the entity in question. Data collection includes understanding cyber exploit details, the initial impact on the entity, and relevant mitigation steps provided to the entity during the information-sharing stage which will leverage operational center partnerships with the government, industry, and international partners to obtain and share situational awareness about incident consequences.

4.1.5 Produce and Share Analysis

Once the vulnerability and consequence cyber-physical integrated analysis is fully completed based on all source information, the analyst will produce and share appropriate analysis data with strategically relevant and actionable information with partners for further feedback and fine-tuning of the analysis.

4.1.6 Provide Situational Awareness

The cyber analyst, in collaboration with other experts will work closely with critical infrastructure owners/operators, Federal agencies, and State and local governments through the local government entities to provide initial information on the threat, risk, and consequence of the cyber incident for situational awareness and data analysis to develop and share actionable mitigation recommendations.

4.2 Mitigation Activities

4.2.1 Identify and Review Physical System Configuration

Based on the sector and the incident, the analyst will identify, understand, and review the entity- or sector-specific infrastructure physical attributes, processes, operational nuances, and mechanical and operational constraints of the systems in question to plan mitigation and response activities. For example, if the incident is related to a natural gas pipeline, the analyst will gather all the relevant operation and physical configuration data, resiliency of the system, and physical operational constraints. The analyst will understand both upstream and downstream infrastructure components such as processing plants, compressor stations, receipt/delivery points, interconnections, and underground storage connectivity, as well as all the pertinent links and nodes for comprehensive mitigation steps to be undertaken by the entity/sector participant experiencing the cyber incident.

4.2.2 Estimate Recovery of the Systems

Based on the understanding of the infrastructure in question and in collaboration with the sector partner and the cyber analyst will develop a timeline for the recovery of the information/control systems. Recovery will be tied to specific steps taken and the timing of the steps taken to support response and recovery plans.

4.2.3 Develop and Implement Courses of Action

Cyber analysts will develop courses of actions in response to the reported cyber incident. These actions may address short-term and long-term responses to mitigate vulnerabilities, weaknesses, or misconfigurations that may be the specific targets of exploits. For example, in the case of a confirmed phishing attack with defined indicators based on analysis, the analyst and the operational centers will collaborate to assess the cost and efficacy of courses of action deemed necessary for the effective development of response and recovery, as appropriate and applicable to the sector-specific infrastructure and operational requirements (e.g., implementing a firewall-blocking rule at the gateway to the business network).

4.3   Response and Recovery

4.3.1   Describe Resiliency of the Infrastructure in Question to Determine Response and Recovery Action Plans

To develop response and recovery action plans, it is vital for the analyst to understand and document the operational resiliency aspects of the sector infrastructure or the individual entity experiencing the cyber exploits on their business and control system networks. This should include discussion with personnel in control systems engineering and operations who can provide insight into the details of how control systems are deployed within the organization, such as the following:

- How are networks typically segregated?
- What remote access connections are generally employed?
- How are high-risk control systems, ESDs, or SISs typically designed?
- How much system downtime can the organization tolerate? How does this downtime compare with the mean repair/recovery time?
- What other processing or communications options can the user access?
- Could a system or security malfunction or unavailability result in injury or death?
- What security countermeasures and mitigation controls are commonly used or in place to respond to a cyber-physical incident?
- How is the security incident response plan put into action for response and recovery of physical assets?
- What type of redundancies and backups are in place to operate the system?
- What workarounds are available for manual control of operation?
- What are the types and availabilities of resources in place (technical/certified personnel) to operate the control system under manual control, and other process plant functions?

4.3.2   Identify Constraints and/or Limitations of the Response and Recovery Action Plans

In collaboration with the sector partners, the cyber analyst documents the relevant operational workarounds and constraints to formulate action plans for response and recovery for efficient and effective response to an incident. An incident response and recovery plan includes a predetermined set of instructions or procedures to detect, respond to, and limit consequences of incidents that target an organization's information and control systems. Response should be measured first and foremost against the service being provided, not just the system that was compromised. The response plan should take into account risk assessments performed initially in collaboration with the sector partner to evaluate both the effect of the attack and the available response options. For example, one possible response is to physically isolate the system under attack. However, this may have such a dire impact on the service that it is dismissed as not viable.[1]

4.3.3 Project Timeframe for Response and Recovery Plans

The development and documentation of response and recovery plans should include appropriate mitigation security controls timeframe (short-term versus long-term) implementation.

4.3.4 Local, State, Regional, and National Consequences

In collaboration with the operation center, the cyber analyst will characterize the incident based on local, State, regional, and national consequences to develop appropriate response and recovery action plans with designations agreed upon and according cyber policy and procedures (catastrophic, major, moderate, and minor) to support the sector entity or infrastructure partner in implementing mitigation and recovery options. Characterizations are as follows (per NIST Special Publication 800-30, with some modification):

- **Catastrophic (High)**—The cyber-physical incident (1) may result in national-level consequences; (2) may result in the very costly loss of major tangible assets or resources or infrastructure; (3) may result in major cascading impacts to other critical infrastructure; (4) may significantly violate, harm, or impede an organization's mission, reputation, or economic interest; and/or (5) may result in human fatality or serious injury.
- **Major (Medium)**—The cyber-physical incident (1) may result in regional-level consequences; (2) may result in the costly loss of tangible assets or resources or some part of infrastructure; (3) may result in some cascading impacts to some other regional critical infrastructure; (4) may violate, harm, or impede an organization's mission, reputation, and interest; or (5) may result in human injury.
- **Moderate (Low)**—The cyber-physical incident (1) may result in State or local-level consequences; (2) may result in the loss of some tangible assets or resources, or a limited part of the infrastructure; (3) may result in limited cascading impacts to some other local-/State-level critical infrastructure; or (4) may noticeably affect an organization's mission, reputation, or interest.

---

[1]Stouffer et al. [1].

4.3.5   Qualitative/Quantitative Likelihood and Consequence of Disruption Event Response

In conducting the cyber-physical incident impact analysis, consideration should be given to the advantages and disadvantages of quantitative versus qualitative assessments. The main advantage of the qualitative impact analysis is that it prioritizes the risks and identifies areas for immediate improvement in addressing vulnerabilities. The disadvantage of the qualitative analysis is that it does not provide specific quantifiable measurements of the magnitude of impacts, thus making a cost-benefit analysis of any recommended controls difficult. The major advantage of a quantitative impact analysis is that it provides a measurement of the impacts' magnitude which can be used in the cost-benefit analysis of recommended controls. The disadvantage is that, depending on the numerical ranges used to express the measurement, the meaning of the quantitative impact analysis may be unclear, requiring the result to be interpreted in a qualitative manner.[2]

Additional factors for consideration to determine the magnitude of impact are as follows:

- **Physical impact**—Physical impacts encompass the set of direct consequences of ICS failure. The potential effects of paramount importance include personal injury and loss of life. Other effects include the loss of property (including data) and potential damage to the environment.
- **Economic impacts**—Economic impacts are a second-order effect of physical impacts ensuing from an ICS incident. Physical impacts could result in repercussions to system operations, which in turn inflict a greater economic loss on the facility or organization. On a larger scale, these effects could negatively affect the local, regional, national, or global economy.
- **Social impacts**—Another second-order effect, the consequence from the loss of national or public confidence in an organization, is many times overlooked. It is, however, a very real target and one that could be accomplished through an ICS incident.[3]

4.3.6   Product Distribution

During the incident analysis, various types/levels of information in various timeframes will be developed for distribution to DHS leadership, as well as Federal and sector partners. The final integrated cyber-physical analysis should take into account audience consideration and actionable information. The current IAC product distribution is as follows:

---

[2] Stoneburner and Goguen [2].

[3] Stouffer et al. [1].

- **Phase 1**—Minimum, mid-level cyber events—Unfinished products (email), requests for information, and bulletins.
- **Phase 2**—Catastrophic—High-level Cyber Events with the following products disseminated: IIAs, IOCs, IRQs, and geospatial bulletins.

4.4 Cyber-Physical Digital Media Analysis

Cyber-physical digital media analysis, also known as computer and network forensics, has many definitions, practices digital media analysis as a means to evaluate a cyber incident. In practice, the digital media analysis is considered the application of science to the identification, collection, examination, and analysis of data while preserving the integrity of the information and maintaining a strict chain of custody for the data. Data refers to distinct pieces of digital information that have been formatted in a specific way. Organizations have an ever-increasing amount of data from many sources. For example, data can be stored or transferred by standard computer systems, networking equipment, computing peripherals, PDAs, consumer electronic devices, and various types of media, among other sources. Because of the variety of data sources, digital media analysis techniques can be used for many purposes, such as internal policy violations, reconstructing computer security incidents, troubleshooting operational problems, and recovering from accidental system damage. The process for performing digital media analysis comprises the following basic phases:

- **Collection**—Identifying, labeling, recording, and acquiring data from the possible sources of relevant data, while following procedures that preserve the integrity of the data.
- **Examination**—Forensically processing collected digital media data using a combination of automated and manual methods, and assessing and extracting data of particular interest, while preserving the integrity of the data.
- **Analysis**—Analyzing the results of the examination, using legally justifiable methods and techniques, to derive useful information that addresses the questions that were the impetus for performing the collection and examination.
- **Reporting**—Reporting the results of the analysis, which may include describing the actions used, explaining how tools and procedures were selected, determining what other actions need to be performed (e.g., examination of additional data sources, securing identified vulnerabilities, improving existing security controls), and providing recommendations for improvement to policies, procedures, tools, and other aspects of the analysis process.[4]

---

[4] Kent et al. [4].

References

1. Stouffer, K., Falco, J., Scarfone, K.: Special publication SP800-82: guide to industrial control systems (ICS) security. NIST (2015). Available at http://csrc.nist.gov/publications/nistpubs/800-82/SP800-82-final.pdf
2. Stoneburner, G., Goguen, A.: Special publication 800-30—risk management guide for information technology systems, NIST (2015, July). Available at http://csrc.nist.gov/publications/nistpubs/800-30/sp800-30.pdf. Accessed 25 Mar 2013
3. Stouffer, K., Falco, J., Scarfone, K.: Special publication SP800-82: guide to industrial control systems (ICS) security. NIST (2015). Available at http://csrc.nist.gov/publications/nistpubs/800-82/SP800-82-final.pdf. Accessed 25 Mar 2013
4. Kent, K., Chevalier, S., Grance, T., Dang, H.: Special publication SP800-86, guide to integrating forensic techniques into incident response. NIST (2006, August). Available at http://csrc.nist.gov/publications/nistpubs/800-86/SP800-86.pdf. Accessed 25 Mar 2013

Chapter 5
Cloud Architecture

Cloud Computing is currently an essential part of network infrastructure. The U.S. National Institute for Standards and Technology has proposed defining cloud computing as a model "for enabling convenient, on-demand network access to a shared pool of configurable computing resources." The cloud structure facilitates the user to access the real-time metadata in a quick and more efficient manner. Data uploads into a centralized data center and is distributed through the network in milliseconds (Fig. 5.1).

5.1 Cloud Service Models

The three basic models with which cloud provides services are, **Software as a Service (SaaS)**, **Platform as a Service (PaaS)**, and **Infrastructure as a Service (IaaS)**.

In **Software as a Service (SaaS)** the cloud provider gives a customer access to applications running in the cloud. The Customer has no control over the infrastructure or even most of the application capabilities but the customer accesses and uses the application.

In **Platform as a Service (PaaS)** the customer has his or her own applications, but the cloud affords the languages and tools for creating them. The customer has no control over the infrastructure that underlies the tools but may have some say in infrastructure configuration.

In **Infrastructure as a Service (IaaS)** the cloud offers processing, storage, networks, and other computing resources that enable customers to run any kind of software. Here, customers can request operating systems, storage, some applications, and some network components.

© Springer Nature Switzerland AG 2021
Fiedelholtz, *The Cyber Security Network Guide*, Studies in Systems, Decision and Control 274, https://doi.org/10.1007/978-3-030-61591-8_5

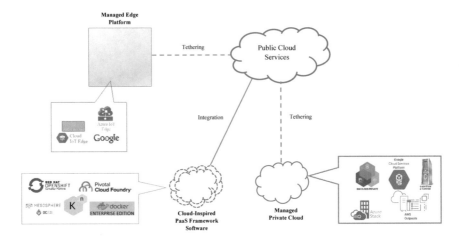

Fig. 5.1 Ubiquitous influence of cloud computing

5.2   Deployment Models

There are many different definitions of clouds, and many ways of describing how clouds are deployed. Often, four basic offerings are described by cloud providers: **private clouds**, **community clouds**, **public clouds**, and **hybrid clouds**.

A **private cloud** has infrastructure that is operated exclusively by and for the organization that owns it, but cloud management may be contracted out to a third party. A **community cloud** is shared by several organization and is usually intended to accomplish a shared goal. For instance, collaborators in a community cloud must agree on its security requirements, policies, and mission. It too, may farm out cloud management to another organization. A **public cloud** available to the general public is owned by an organization that sells cloud services. A **hybrid cloud** is composed of two or more types of clouds, connected by technology that enables data and applications to be moved around the infrastructure to balance loads among clouds.

Thus, cloud software is not business as usual. It must provide services without anchoring in a particular location. It must also be highly modular, with low coupling and easy interoperability all characteristics of good code.

Two of the major cloud models that are most used by the government and private sector are the Amazon Web Cloud Services (AWS) and Azure Microsoft Web Services.

5.3   Amazon Web Services (AWS) Cloud Models

Amazon Web Services (AWS) Cloud functions as a micro service is an approach to data application development in which a large application is built as a suite of modular components over the internet services. The AWS Cloud network is established to enhance the efficiency and effectiveness of cyber data being transferred over the internet.

The government cloud data in the AWS platform is encrypted with a public and private key is used to secure Confidentiality, Integrity, and Availability (CIA).

5.4 Azure Microsoft Web Services Cloud Models

Azure Cloud Platform is created by Microsoft and also utilized in a government ecosystem. It is a virtualized platform connected to a computer CPU to transmit data throughout the network.

In many government agencies, the Azure Cloud is utilized for user authority and authentication purposes to ensure Confidentiality, Integrity and Availability (CIA).

The Azure hypervisor is utilized to initiate the server to transmit the data over the Application Programming Interface (API) to send the authentication response.

Chapter 6
Lessons Learned

Once the cyber-physical analysis and incident response and recovery action and plans are communicated to the sectors or entities in question, the cyber operational agencies, sector agencies, and the sector or entity involved should review the entire incident analysis process to determine what worked and identify areas of weakness to improve the process for the next incident response and recovery analysis.

The lessons learned process should strive to identify shortcomings of the process that walks through the D Cyber-Physical Mapping Framework Analysis Process Matrix depicted in Appendix G (i.e., pre-incident planning and analysis, incident detection and characterization, vulnerability/consequence analysis, incident response and recovery, roles, responsibilities, collaboration, support, various report outputs, and training). The lessons learned should provide guidance for the additional need for training for all the personnel involved in the cyber-physical incident analysis.

The lessons learned should also focus on collecting data, sharing information, providing guidance and assisting the sector or entity in preventing future attacks, preventing or limiting disruption if they do occur, and creating early visibility of such attacks through enhanced awareness, security monitoring, and training.

Other key components of an effective post-event analysis program include the following:

- Identify what transpired and the sequence of events;
- Understand the causes of events;
- Understand the vulnerabilities that were exploited;
- Identify and ensure timely implementation of corrective actions;
- Develop and disseminate recommendations and valuable lessons learned to the Federal partners, and industry to enhance operational performance and avoid repeat events;
- Develop the capability for integrating risk analysis into the event analysis process; and

© Springer Nature Switzerland AG 2021

Fiedelholtz, *The Cyber Security Network Guide*, Studies in Systems, Decision and Control 274, https://doi.org/10.1007/978-3-030-61591-8_6

- Feed forward key results to facilitate enhancements in and support of the various Federal programs and initiatives (e.g., performance metrics, standards, compliance monitoring and enforcement, training, and education).[1]

Reference

1. NERC (North American Electric Reliability Corporation): Cyber attack task force—final report. Available at http://www.nerc.com/docs/cip/catf/12-CATF_Final_Report_BOT_clean_Mar_26_2012-Board%20Accepted%200521.pdf (2012). Accessed 25 Mar 2013

---

[1]NERC [1].

Appendix A
Cyber Network Hardware and Software Operating Procedure (SOP)

© Springer Nature Switzerland AG 2021
Fiedelholtz, *The Cyber Security Network Guide*, Studies in Systems,
Decision and Control 274, https://doi.org/10.1007/978-3-030-61591-8

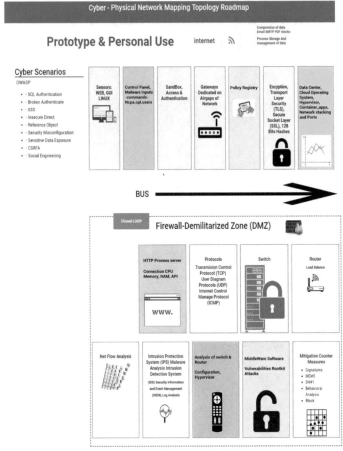

OWASI- Open Web Application Security Project- 501(c) (3), NIST 800-53 Security and Privacy Controls for Information Systems and Organization- August 2017
NOTE: Colors depict major vulnerabilities on the network
Designed by Media Graphics Lab   Phone: (301) 460-7678   Email: mediagraphicslab@gmail.com

Term	Description
Sensor	Sensors typically are small inexpensive and often battery powered meaning that issues of very low power consumption and limited process capability
Control panel	Network Control Panel enables us to change various hardware and software features. For example, System and Security, Network and Internet, Hardware and Software Programs, User Accounts
SandBox	Applets and servlets run in a protected area called the sandbox, which provides extensive security controls to prevent them from accessing unauthorized resources or damaging the hardware OO or file system

(continued)

(continued)

Term	Description
Gateways	Nodes connecting two or more networks or network segments that might be physically implemented as workstations, servers, or routers. (Layer 5)
Policy registry	Establish hardware and software policies for governing network features
Encryption	The act of applying a transforming function to data, with the intention that only the receiver of the data will be able to read it (after applying the inverse function, *decryption*). Encryption generally depends on either a secret shared by the sender and receiver or a public/private key pair
Cloud	A specific way of organizing computer resources for maximum availability and accessibility and minimal complexity in the interfaces including front end web based interfaces and a large collection of computing and data resources (collectively called back end resources)
Bus	A communication protocol used by all devices attached to a bus that governs the formats, content and timing of data, memory address and control messages sent across the bus
Firewall—Demilitarized Zone (DMZ)	The DMZ is used to hold services such as DNS and email servers that need to be accessible to the outside. Both the internal network and outside world can access the DMZ but hosts in the DMZ cannot access internal network. Therefore, an adversary who succeeds in compromising a host in the exposed DMZ still cannot access the DMZ
HTTP	HTTP is an application protocol that is used to retrieve webpages from remote servers. May efficient application programs that is web clients like Internet Explorer, Chrome, Firefox, and Safari provide users with different look and feel but all of them use the same HTTP protocol to communicate web servers over the Internet
Protocols	A specification of an interface between modules running on different machines, as well the communication service that those modules implement. The terms is also used to refer to an implementation of the module that meets this specification. To distinguish between these two uses, the interface is often called a *protocol specification*
Switch	A network node that forwards packets from inputs to outputs based on header information in each packet. Differs from a *router* mainly in that it typically does not interconnect networks of different types
Router	A network node connected to two or more networks that forwards packets from one network to another. Contrast with *bridge*, *repeater*, and *switch*

(continued)

(continued)

Term	Description
Netflow analysis	Strictly speaking, a *flow* is a series of packets that share the same source and destination IP addresses, source and destination ports, and IP protocol. This is also called a *five-tuple IP flow*. The word *flow* is also sometimes used to mean an aggregate of individual flows. A *flow record* is a summary of information about a flow, recording which hosts communicated with which other hosts, when this communication occurred, how the traffic was transmitted, and other basic information about the network conversation. A flow analysis system collects flow information and gives you a system to search, filter, and print flow information. Flow records summarize every connection on your network
Intrusion detection system; Intrusion protection system	Related to firewalls are systems known as *intrusion detection systems* (IDS) and *intrusion protection systems* (IPS). These systems try to look for anomalous activity such as an unusually large amount of traffic targeting a given host or port number, for example, and generate alarms for network managers or perhaps even take direct action to limit a possible attack. While there are commercial products in this space today, it is still a developing field
Middleware	System software that glues together parts of a client/server or multitier application and enables clients and servers or distributed components to locate and communicate with one another
Mitigation	Cyber operational agencies promulgate countermeasures to mitigate suspicious IP addresses. (*see mitigation diagram*)

Appendix B
Cyber-Physical Mapping Framework Analysis Process Matrix

Pre-Incident Planning and Analysis	Incident Detection and Characterization	Vulnerability/Consequence Analysis	Incident Response and Recovery
Steady-state Monitoring – Continuous Monitoring	Detection	Vulnerability/Consequence Analysis	Mitigation Activities
Initial Watch and Warning	Threat Analysis	Dependency/Interdependency Analysis	Response and Recovery
	Malware Analysis	Analysis Reporting	Cyber-Physical Digital Media Analysis
	Information Sharing		
	Structured Threat Information eXpression (STIX)		

Information Sharing

Lessons Learned

Appendix C
OWASP Top Ten Cyber Attacks[1]

1. **Targeted attacks**—Attacks that are tailored to penetrate a particular sector organization or a company focused on gathering sensitive data that has a monetary value in the underground market.
2. **Hactivism**—Attacks that include the DDoS attacks by various perpetrators. Examples of DDoS attacks in 2012 include those launched by Anonymous on government Web sites in Poland, following the government's announcement that it would support the Anti-Counterfeiting Trade Agreement; the hacking of the official Formula 1 Web site in protest against the treatment of anti-government protesters in Bahrain; the hacking of various oil companies in protest against drilling in the Arctic; the attack on Saudi Aramco; and the hacking of the French Euro millions Web site in a protest against gambling. Another example of recent DDoS took place in March and into April 2013; Mt. Gox, the largest Bitcoin Exchange in the world, battled a continued DDoS attack in an attempt to destabilize the currency and/or profit from it.
3. **Cyber espionage and warfare**—Stuxnet pioneered the use of highly sophisticated malware for targeted attacks on key production facilities. However, while such attacks are not commonplace, it is now clear that Stuxnet was not an isolated incident. The global community is entering an era of cold "cyber-war," where nations have the ability to fight each other unconstrained by the limitations of conventional real-world warfare. Looking ahead, we can expect more countries to develop cyber weapons.
4. **Big Brother watching even more**—This will include using technology to monitor the activities of those suspected of criminal activities. This is not a new issue: consider the controversy surrounding "Magic Lantern" and the "Bundestrojan." More recently, there has been debate around reports that a U.K. company offered the "Finfisher" monitoring software to the previous Egyptian government and reports that the Indian government asked firms (including Apple, Nokia, and RIM) for secret access to mobile devices. The use of legal surveillance tools has wider implications for privacy and civil liberties. Further,

---

[1]Forbes [1].

© Springer Nature Switzerland AG 2021
Fiedelholtz, *The Cyber Security Network Guide*, Studies in Systems,
Decision and Control 274, https://doi.org/10.1007/978-3-030-61591-8

as law enforcement agencies and governments try to get one step ahead of the criminals, it is likely that the use of such tools, and the debate surrounding their use, will continue.

5. **Malware exploits**—Malware exploits are common to both the wired and wireless networks. Today, the wide use of mobile devices, while offering huge benefits to a business, also increases its risk from malware exploits that allow the perpetrator access to the company's enterprise business and control system networks. Cloud data can be accessed from devices that may not be as secure as traditional endpoint devices. When the same device is used for both personal and business tasks, risk increases still further. In December 2012, Iran reported new Stuxnet-like Trojan attack targeting Iran's oil refinery control systems. A Trojan is a type of malware designed to give remote control of one computer to another in order to inflict damage or steal information. Famous Trojans have included Zeus and Netbus. Preliminary data suggested that files on several computers were overwritten with garbage code, after which the hard disks on the targeted systems were wiped clean by a malicious program.

6. **Privacy breaches**—The value of personal data to cybercriminals and legitimate businesses will only grow in the future, and with it the potential threat to privacy increases.

7. **Cyber extortion**—The year 2012 saw growing numbers of ransomware Trojans designed to extort money from their victims, either by encrypting data on the disk or by blocking access to the system. Such attacks are easy to develop and, as with phishing attacks, there seems to be no shortage of potential victims. Until fairly recently, this type of cybercrime was confined largely to Russia and other former Soviet countries. However, they have now become a worldwide phenomenon, although sometimes with slightly different modus operandi. In Russia, for example, Trojans that block access to the system often claim to have identified unlicensed software on the victim's computer and ask for a payment.[2]

8. **Apple MAC and iPad OS attacks**—Attacks on the Mac OS have grown steadily over the last 2 years; it would be naive of anyone using a Mac to imagine that they could not become the victim of cybercrime. It is not only generalized attacks, such as the 700,000-strong Flashfake botnet, that pose a threat; there have also been targeted attacks on specific groups, or individuals, known to use Macs. The threat to Macs is real and is likely keep growing. In addition, recent events have identified exploits of Apple iPad OS; iPad use is becoming popular for normal business uses such as accessing emails and corporate data.

9. **Attack on Android OS**—Mobile malware has exploded in the last 18 months because of the popularity and use of smartphones by employees and business executives. The lion's share of it targets Android-based devices—more than 90% is aimed at this OS. The appearance of the "Find and Call" app earlier this year has shown that it is possible for undesirable apps to slip through the net. However it is likely that, for the time being at least, Android will remain the chief focus of cyber criminals. The key significance of the Find and Call app

---

[2]Croucher [2].

lies in the issue of privacy, data leakage, and the potential damage to a person's reputation: this app was designed to upload someone's phonebook to a remote server and use it to send Systems Management Server (SMS) spam.

10. **Java exploits**—One of the key methods used by cyber criminals to install malware on a computer is to exploit unpatched vulnerabilities in applications. This relies on the existence of vulnerabilities and the failure of individuals or businesses to patch their applications. Java vulnerabilities currently account for more than 50% of attacks, while Adobe Reader accounts for another 25%. Cyber criminals will continue to exploit Java in the year ahead. It is likely that Adobe Reader will also continue to be used by cyber criminals, but probably less so because the latest versions provide an automatic update mechanism.

Appendix D
Structured Threat Information EXpression (STIX™)

D.1   STIX™

STIX™ is a structured language for cyber threat intelligence information used by Federal entities for the capture, characterization, and communications of standardized cyber threat information improving consistency, efficiency, interoperability, and overall situational awareness. A variety of high-level cybersecurity use cases rely on cyber situational awareness information provided by STIX™, TAXII™, and CybOX™. Cyber situational awareness is outlined below:

(I)   Cyber situational awareness

    (A)   Share techniques for incident response—Leverage STIX™, TAXII™, and CybOX™ cyber observables and techniques to provide situational awareness and response. Provide guidance on incident management information utilizing attack patterns and malware characterizations from the analysis.

    (B)   Share identified attack indicators—Leverage STIX™, TAXII™, and CybOX™ cyber observables from attack detection data to identify and share malicious activity, attack patterns, and behavior.

In addition, STIX™ provides a unifying architecture tying together a diverse set of cyber threat information including the following:

- Cyber observables,
- Indicators,
- Incidents,
- Adversary tactics, techniques, and procedures (including attack patterns, malware, exploits, kill chains, tools, infrastructure, and victim targeting),
- Exploit targets (e.g., vulnerabilities, weaknesses, or configurations),
- Courses of action (e.g., incident response or vulnerability/weakness remedies or mitigations),
- Cyber-attack campaigns, and
- Cyber threat actors.

© Springer Nature Switzerland AG 2021
Fiedelholtz, *The Cyber Security Network Guide*, Studies in Systems,
Decision and Control 274, https://doi.org/10.1007/978-3-030-61591-8

STIX™ directly leverages the following constituent schema:

- **Cyber Observable eXpression (CybOX™)**—CybOX™ is a standardized schema for the specification, capture, characterization, and communication of events or stateful properties that are observable in the operational domain. CybOX™ covers most of the framework process such as detection, threat analysis, malware analysis, and incident response and sharing activities. For more information on the attributes of CybOX™, see Appendix D.
- **Common Attack Pattern Enumeration and Classification (CAPEC™)**—CAPEC™ is a publicly available, community-developed list of common attack patterns, along with a comprehensive schema and classification taxonomy. Attack patterns are descriptions of common methods for exploiting software systems. They derive from the concept of design patterns applied in a destructive rather than constructive context and are generated from in-depth analysis of specific real-world exploit examples.
- **Malware Attribute Enumeration and Characterization (MAEC™)**—MAEC is a standardized language for encoding and communicating high-fidelity information about malware based upon attributes such as behaviors, artifacts, and attack patterns. MAEC™ aims to improve human-to-human, human-to-tool, tool-to-tool, and tool-to-human communication about malware; reduce potential duplication of malware analysis efforts by researchers; and allow for the faster development of countermeasures by enabling the ability to leverage responses to previously observed malware instances.
- **Common Vulnerability Reporting Framework**—The Industry Consortium for Advancement of Security on the Internet Common Vulnerability Reporting Framework (CVRF) is an Extensible Markup Language (XML)-based language that enables different stakeholders across different organizations to share critical security-related information in a single format, speeding up information exchange and digestion. CVRF is a common and consistent framework for exchanging not just vulnerability information, but any security-related documentation.[3]

STIX™ and CybOX are used in conjunction with TAXII™ that allows sharing of actionable cyber threat information across organization. The following describes the attributes of CybOX™ and TAXII™:

D.2 CybOX™

CybOX™ is a standardized schema for the specification, capture, characterization, and communication of events or stateful properties that are observable in the operational domain. CybOX™ covers most framework processes such as detection, threat analysis, malware analysis, and incident response and sharing activities.

---

[3]The MITRE Corporation [3].

CybOX™ was developed by MITRE with the sponsorship of Federal organizations. The concept of observable events or properties in the operational cyber realm is a central underlying element of many of the different activities involved in cybersecurity. There currently exists no uniform standard mechanism for specifying, capturing, characterizing, or communicating these cyber observables. Each activity area, each use case, and often each supporting tool vendor uses its own unique approach, which inhibits consistency, efficiency, interoperability, and overall situational awareness. CybOX™ is targeted to support a wide range of relevant cybersecurity domains including the following:

- Threat assessment and characterization (detailed attack patterns),
- Malware characterization,
- Determine operational event management,
- Conduct log analysis,
- Conduct digital forensics,
- Cyber situational awareness,
- Incident response, and
- Indicator sharing.

Through utilization of the standardized CybOX™ language, relevant observable events or properties can be captured and shared, defined in indicators and rules, or used to reveal the appropriate portions of attack patterns and malware profiles in order to tie the logical pattern constructs to real-world evidence of their occurrence or presence for attack detection and characterization. Incident response and management can then take advantage of all of these capabilities to investigate occurring incidents, improve overall situational awareness, and improve future attack detection, prevention, and response.

Threat assessment and characterization (detailed attack patterns)—A wide variety of high-level cybersecurity use cases rely on such information, including event management/logging, malware characterization, intrusion detection, incident response/management, and attack pattern characterization. CybOX™ provides a common mechanism (structure and content) for addressing cyber observables across and among this full range of use cases, improving consistency, efficiency, interoperability, and overall situational awareness. The following are tasks performed by CybOX™:

- **Identify malware characterization**—Currently, the AV industry categorizes malware based on main malicious activities (e.g., virus, worm, spyware, fake AV, and adware). In an attempt to account for expanding malware behavioral variety, several major AV companies adopted more detailed, tree-based malware classification. The recent discoveries of sophisticated malware including Stuxnet and Flame demonstrated that the evolution of the mainstream malware techniques to include stealthy, precise cyber weapons aimed to disrupt critical infrastructure and exfiltrate sensitive information. This has caused researchers to develop and apply a novel approach for building soft clusters that expose behavioral commonalities characterized as component traits. The idea is to identify malware in terms of component-based malware cluster, behavior mapping, and malware relationship.[4]

- **Determine operational event management**—Operational event management logged data from security event systems is monitored and compiled from network and application firewalls, network and host IDS/IPS, access controls, sniffers, and Unified Threat Management systems (UTM), switch, router, load balancer, OS, server, badge reader, custom or legacy application, and many other IT systems across the enterprise.[5] From this data, CybOX™ allows the compilation of the observable events in the operational domain.

- **Determine Logging**—Logging records are the only evidence of a successful attack. Many organizations keep audit records for compliance purposes, but attackers rely on the fact that such organizations rarely look at the audit logs, so they do not know that their systems have been compromised. Because of poor or nonexistent log analysis processes, attackers sometimes control victim machines for months or years without anyone in the target organization knowing, even though the evidence of the attack has been recorded in unexamined log files. Deficiencies in security logging and analysis allow attackers to hide their location, malicious software used for remote control, and activities on victim machines. Even if the victims know that their systems have been compromised, without protected and complete logging records they are blind to the details of the attack and to subsequent actions taken by the attackers. Without solid audit logs, an attack may go unnoticed indefinitely and the particular damages done may be irreversible.[6]

- **Apply digital forensics**—Through utilization of the standardized CybOX™ language, collect digital forensics relevant to observable events or properties on the appropriate portions of attack patterns and malware profiles from CybOX™ techniques in order to tie the logical pattern constructs to real-world evidence of their occurrence or presence for attack detection and characterization.[7]

D.3 TAXII™

Defines a set of services and message exchanges that, when implemented, enable sharing of actionable cyber threat information across organization and product/service boundaries. TAXII™, through its member specifications, defines concepts, protocols, and message exchanges to exchange cyber threat information

---

[4] Yavvari et al. [4].

[5] Butler [5].

[6] SANS Institute [6].

[7] The MITRE Corporation [7].

for the detection, prevention, and mitigation of cyber threats. TAXII™ is the preferred method of exchanging information represented using the STIX™ language, enabling organizations to share structured cyber threat information in a secure and automated manner.[8]

TAXII™ roles (producer and consumer) are used to denote participants in TAXII™ according to their high-level objectives in the use of TAXII™ services:

- **Producer**—An entity (e.g., a person, organization, or agency) that is the source of structured cyber threat information.
- **Consumer**—An entity that is the recipient of structured cyber threat information. (Note that these roles are not mutually exclusive; one entity might be both a consumer and a producer of structured cyber threat information.)
- **TAXII™ Functional Units**—TAXII™ functional units represent discrete sets of functionality required to support TAXII™. A single software application could encompass multiple functional units or multiple applications could cooperate to serve as a single functional unit. A functional unit simply represents some component with a well-defined role in TAXII™.
- **TAXII™ Message Handler (TMH)**—A functional unit that produces and consumes TAXII™ messages. The TMH is responsible for parsing inbound TAXII™ messages and constructing outbound TAXII™ messages in conformance with one or more TAXII™ message binding specifications. A TMH interacts with the TAXII™ Transfer Agent (defined below), which handles the details required to transmit those messages over the network. The TAXII™ back-end interacts with the TMH to turn the information from the back-end into TAXII™ Messages and to perform activities based on the TAXII™ messages that the TMH receives.
- **TAXII™ Transfer Agent (TTA)**—A network-connected functional unit that sends and/or receives TAXII™ messages. A TTA interacts with other TTAs over the network and handles the details of the protocol requirements from one or more TAXII™ protocol binding specifications. A TTA provides TAXII™ messages to a TMH, allowing the TMH to be agnostic to the utilized network protocol. By the same token, the TTA can be agnostic as to the content of TAXII™ Messages, leaving the handling of this information to the TMH.
- **TAXII™ Back-end**—A term that covers all functional units in the TAXII™ architecture other than the TTA and the TMH. This could cover data storage, subscription management, access control decisions, filtering of content prior to dissemination, and other activities. The TAXII™ specifications provide no requirements on how capabilities are implemented in a TAXII™ back-end beyond noting that TAXII™ back-ends need to be able to interact with a TMH. Individual implementers and organizations can decide which TAXII™ back-end capabilities are necessary given the TAXII™ services they wish to support and how they wish to provide this support.
- **TAXII™ Architecture**—A term that covers all functional units of a single producer or consumer's infrastructure that provide and/or utilize TAXII™

---

[8]The MITRE Corporation [8].

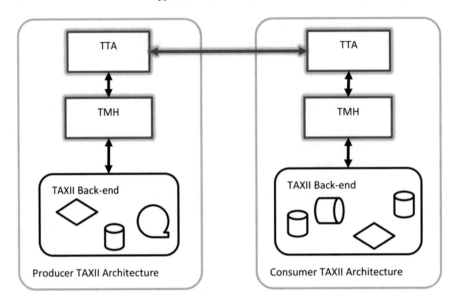

Fig. D.1 The interaction of TAXII™ functional units (Ibid.)

services. A TAXII™ Architecture includes a TTA, a TMH, and a TAXII™ back-end. As noted above, implementation details of a TAXII™ back-end are outside of the scope of the TAXII™ specifications. Figure D.1 shows the interaction of TAXII™ functional units.[9]

---

[9]Davidson and Schmidt [9].

Appendix E
Open Systems Interconnection (OSI) Reference Model

The OSI Reference Model has seven layers: application, presentation, session, transport, network, data link, and physical. Understanding severity of the cyber incident and exploit requires understanding of these seven layers. The OSI Reference Model defines Internet working in terms of a vertical stack of seven layers. The upper layers of the OSI Reference Model represent software that implements network services, such as encryption and connection management. The lower layers of the OSI Reference Model implement more primitive, hardware-oriented functions like routing, addressing, and flow control (Fig. E.1).

It is a fundamental rule that higher layers cannot be secured without the lower layers also being secured. For example, currently known/familiar threats at lower levels of the OSI stack include address resolution protocol (ARP) spoofing man-in-the-middle (MITM) attacks at layer two, and physical layer attacks such as passive optical taps or the interception of wireless network signals by attackers. As noted by SANS Institute in their Top 20 Security Risks report, nearly half of the 4396 total vulnerabilities reported in SANS @RISK data from November 2006 to October 2007

Fig. E.1 Seven layers of OSI reference model (University of Washington [10])

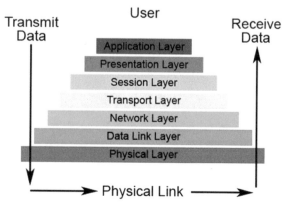

Fiedelholtz, *The Cyber Security Network Guide*, Studies in Systems, Decision and Control 274, https://doi.org/10.1007/978-3-030-61591-8

relate to Web application vulnerabilities such as Structured Query Language (SQL) Injection attacks, cross-site scripting, cross-site request forgeries, and personal home page (PHP) remote file inclusions. Arguably, proper application of encryption to data in transit and data at rest, along with improved application development practices to eliminate issues such as like SQL injection attacks, should largely mitigate these risks, and yet that is not the case. Below are the seven layers described and explained, beginning with the "lowest" in the hierarchy (the physical) and proceeding to the "highest" (the application).

E.1 Physical Layer

The physical layer is the lowest layer of the OSI Reference Model, which is concerned with the transmission and reception of the unstructured raw bit stream over a physical medium. It describes the electrical/optical, mechanical, and functional interfaces to the physical medium, and it carries the signals for all of the higher layers. The physical layer defines the electrical, mechanical, procedural, and functional specifications for activating, maintaining, and deactivating the physical link between communicating network systems. Physical layer specifications define characteristics such as media, voltage levels, timing of voltage changes, physical data rates, maximum transmission distances, and physical connectors.

This layer ensures that a bit sent on one side of the network is received correctly on the other side. Data travel from the application layer of the sender, down through the levels, across the nodes of the network service, and up through the levels of the receiver. To keep track of the transmission, each layer "wraps" the preceding layer's data and header with its own header. A small chunk of data will be transmitted with multiple layer headers attached to it. On the receiving end, each layer strips off the header that corresponds to its respective level.

Kinetic (physical) destruction and intentional attempts at electromagnetic destruction of network assets via high-power microwave weapons, or high-altitude electromagnetic pulse effects, are examples of the threats at the physical layer of hardware.[10] Another example of physical layer-1 vulnerability is a computer's need for electrical power. A physical exploit would be to pull the power cord out of the wall. Mitigation to prevent the exploit would be to house the power cords in a locked enclosure. A countermeasure would be to connect all the computers to an uninterruptible power supply (UPS) device.[11]

---

[10]University of Oregon [11].

[11]TechExams.net [12].

E.2 Data Link Layer

The data link layer provides error-free reliable transfer of data across a physical network. This layer deals with issues such as flow regulation, error detection and control, and frames. This layer has the important task of creating and managing what frames are sent out on the network. The network data frame, or packet, is made up of checksum, source address, destination address, and the data itself. The largest packet size that can be sent defines the maximum transmission unit. Different data link layer specifications define different network and protocol characteristics, including physical addressing, network topology, error notification, sequencing of frames, and flow control. Physical addressing (as opposed to network addressing) defines how devices are addressed at the data link layer.

Network topology consists of the data link layer specifications that often define how devices are to be physically connected, such as in a bus or a ring topology. Error notification alerts upper-layer protocols that a transmission error has occurred, and the sequencing of data frames reorders frames that are transmitted out of sequence. Finally, flow control moderates the transmission of data so that the receiving device is not overwhelmed with more traffic than it can handle at one time. The protocols used in the data link layer are serial line Internet or interface protocol (SLIP), point-to-point protocol (PPP), MTU, and compressed serial line Internet or interface protocol (CSLP). Logical link control (LLC) defines how data are transferred over the cable and provides data link service to the higher layers. Medium access control (MAC) defines who can use the network when multiple computers are trying to access it simultaneously (e.g., token passing, Ethernet).

E.3 Network Layer

The network layer controls the operation of the subnet, deciding which physical path the data should take based on network conditions, priority of service, and other factors. Figure E.2 shows the test packets captured under the Wireshark tool.

The network layer is responsible for the routing of data (packets) through the network and handles the addressing and delivery of data. This layer provides for congestion control, accounting information for the network, routing, addressing, and several other functions. IP is a good example of a network layer protocol.

Network address range	Default mask
10.0.0.0 - 10.255.255.255	255.0.0.0
172.16.0.0 - 172.31.255.255	255.240.0.0
192.168.0.0 - 192.168.255.255	255.255.0.0

Fig. E.2 Private networks default subnet masks (About.com [13])

- **Communications Subnet**—The network layer software must build headers so
  that the network layer software residing in the subnet intermediate systems can
  recognize them and use them to route data to the destination address. This layer
  relieves the upper layers of the need to know anything about the data transmission
  and intermediate switching technologies used to connect systems. It establishes,
  maintains, and terminates connections across the intervening communications
  facility (one or several intermediate systems in the communication subnet). In
  the network layer and the layers below, peer protocols exist between a node and
  its immediate neighbor, but the neighbor may be a node through which data
  are routed, not the destination station. The source and destination stations may
  be separated by many intermediate systems. Subnetting works by applying the
  concept of extended network addresses to individual computer (and other network
  device) addresses. An extended network address includes both a network address
  and additional bits that represent the subnet number. The governing bodies that
  administer IP have reserved certain networks for internal uses. In general, intranets
  utilizing these networks gain more control over managing their IP configuration
  and Internet access. An example of the default subnet masks associated with these
  private networks is listed in Fig. E.3.

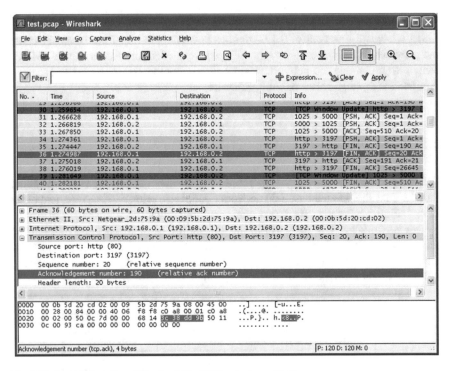

Fig. E.3 Test Packets from Wireshark Tool (Wireshark.org [14])

E.4 Transport Layer

Transport layer is responsible for reliable transmission of data and service specification between hosts. The major responsibility of this layer is data integrity, ensuring that data transmitted between hosts is reliable and timely. Upper-layer datagrams are broken down into network-sized datagrams, if needed, and then implemented using appropriate transmission control. The transport layer creates one or more network connections, depending on conditions. This layer also handles what type of connection will be created. Two major transport protocols are the TCP and the User Datagram Protocol (UDP). The Internet Control Message Protocol (ICMP) is an integral part of any TCP/IP implementation. The transport layer ensures that messages are delivered error-free, in sequence, and with no losses or duplications. It relieves the higher layer protocols from any concern with the transfer of data between them and their peers.

In any typical attack scenario, the attacker will first engage in some reconnaissance and scanning activities using open source automatic tools such as Network Mapper (Nmap) and Superscan in order to better understand the environment of the target; gather information about the target so as to plan the attack approach; and employ the right techniques and tools for the subsequent attack phases. One of the most common (albeit noisy) and most well-understood techniques for discovering the range of hosts that are alive in the target's environment is to perform a ICMP sweep of the entire target's network range. An ICMP sweep involves essentially sending a series of ICMP request packets to the target network range and inferring from the list of ICMP replies whether certain hosts are alive and connected to the target's network for further probing.[12]

- **End-to-end layers**—Unlike the lower "subnet" layers whose protocol is between immediately adjacent nodes, the transport layer and the layers above are true "source to destination" or end-to-end layers, and are not concerned with the details of the underlying communications facility. Transport layer software (and software above it) on the source station carries on a conversation with similar software on the destination station by using message headers and control messages.

E.5 Session Layer

The session layer coordinates dialogue, session, and connection between devices over the network. This layer manages communications between connected sessions. Examples of this layer are token management (the session layer manages who has the token) and network time synchronization. The session layer establishes, manages, and terminates communication sessions by allowing two application processes on different machines to establish, use, and terminate a connection, called a session. The session timeout defines action window time for a user that also represents a delay

---

[12]SANS Institute [15].

during which an attacker can try to steal and use an existing user session. The best practice to protect from exploits is to keep the session timeout to the minimal value possible based on the application.[13]

Communication sessions consist of service requests and service responses that occur between applications located in different network devices and perform the functions that allow these processes to communicate over the network, performing security, name recognition, logging, and so on. These requests and responses are coordinated by protocols implemented at the session layer. Some examples of session-layer implementations include Zone Information Protocol (ZIP), the AppleTalk protocol that coordinates the name binding process, and Session Control Protocol (SCP), the Decent Phase IV session layer protocol.

E.6   Presentation Layer

The presentation layer formats the data to be presented to the application layer in an ordered and meaningful manner. It can be viewed as the translator for the network. This layer's primary function is the syntax and semantics of the data transmission. It converts local host computer data representations into a standard network format for transmission on the network. On the receiving side, it changes the network format into the appropriate host computer's format so that data can be utilized independent of the host computer. American Standard Code for Information Interchange (ASCII) and Extended Binary Coded Decimal Interchange Code (EBCDIC) conversions, cryptography, and the like are handled here.

The presentation layer provides a variety of coding and conversion functions that are applied to application layer data. These functions ensure that information sent from the application layer of one system would be readable by the application layer of another system. Examples of presentation layer coding and conversion schemes include common data representation formats, conversion of character representation formats, common data compression schemes, and common data encryption schemes.

Common data representation formats, or the use of standard image, sound, and video formats (e.g., QuickTime and MPEG [Motion Picture Experts Group], JPEG [Joint Photographic Experts Group], GIF [Graphics Interchange Format], and TIFF [Tagged Image File Format]) allow the interchange of application data between different types of computer systems. Using different text and data representations, such as EBCDIC and ASCII, uses conversion schemes to exchange information with systems. Standard data compression schemes allow data that are compressed or encrypted at the source device to be properly decompressed or deciphered at the destination.

---

[13]OWASP (Open Web Application Security Project) [16].

E.7   Application Layer

The application layer serves as the window for users and application processes to access network services. This layer is the main interface through which the user interacts with the application and therefore the network. The application layer is the most exploitable layer, where the application using the network resides and makes use of applications such as Web browsing, emails, instant messaging, file transfer systems, and remote login systems. This layer is vulnerable primarily to data theft, cross-site scripting, SQL injection, buffer overflows, and numerous other attack methods.

The application layer is the OSI layer closest to the end user, which means that both the OSI application layer and the user interact directly with the software application. This layer interacts with software applications that implement a communicating component. Application layer functions typically include identifying communication partners, determining resource availability, and synchronizing communication. When identifying communication partners, the application layer determines the identity and availability of communication partners for an application with data to transmit. When determining resource availability, the application layer must decide whether sufficient network resources for the requested communication exist. In synchronizing communication, all communication between applications requires cooperation that the application layer manages. Some examples of application layer implementations include Telnet, FTP, and Simple Mail Transfer Protocol (SMTP).[14,15]

E.8   The User

While users are not an explicit layer of the OSI Reference Model, they represent a unique vulnerability. Users are vulnerable to a wide variety of social engineering attacks ranging from deception to stealing of credentials and phishing. Because system integrity can be undercut by users volunteering their passwords, additional research into human factors is needed to better understand how to keep human participants in complex security systems from serving as the "weakest link."

Industry recognizes that access control is vital to protect systems and people from exploits. Secure Shell Hash (SSH) and Secure Socket Layer (SSL)/Transport Layer Security (TLS) encryption along with two-factor authentication (the use of both something the user know, such as a password, and something the user has, such as a hardware cryptographic token), should largely make technical credential capture attempts a futile exercise. However, end-to-end strong encryption and two-factor authentication is still the exception rather than the rule, primarily because of economic and ease-of-use issues.

---

[14]Microsoft [17].

[15]TutorialsWeb.com [18].

Appendix F
Cybersecurity Tools[a]

Security tool	Function	Security information
ArcSight	Provides a suite of tools SIEM	A log analyzer and correlation engine designed to sift out important network events
BackTrack	A LINUX OS	Has a variety of security and forensic tools for penetration testing. For example, the toolset includes network mapping, information gathering, vulnerability identification, Web application analysis, digital forensics, and reverse engineering
Cain and Abel	A security tool that supports UNIX and Windows platforms and scans for passwords for recovery and or decrypting	Can recover passwords by sniffing the network, cracking encrypted passwords using dictionary, brute-force, and cryptanalysis attacks; recording voice-over IP conversations; decoding scrambled passwords; revealing password boxes; uncovering cached passwords; and analyzing routing protocols

(continued)

© Springer Nature Switzerland AG 2021
Fiedelholtz, *The Cyber Security Network Guide*, Studies in Systems,
Decision and Control 274, https://doi.org/10.1007/978-3-030-61591-8

(continued)

Security tool	Function	Security information
Kismet	A console-based 802.11 layer-2 wireless network detector, sniffer, and IDS	Identifies networks by passively sniffing (as opposed to more active tools such as NetStumbler), and can even de-cloak hidden (non-beaconing) networks. It can automatically detect network IP blocks by sniffing TCP, UDP, ARP, and dynamic host configuration protocol (DHCP) packets, log traffic in Wireshark/tcpdump-compatible format, and even plot detected networks and estimated ranges on downloaded maps
Metasploit	An advanced open-source platform for developing, testing, and using exploit code	Generally used to develop and test exploits. The tool comes with ready exploits for use by attackers. The tool can also allow the attacker to explore the exploit code to understand the malware function
Nessus	A popular vulnerability scanner for UNIX systems	Features include remote and local (authenticated) security checks, a client/server architecture with a Web-based interface, and an embedded scripting language for writing plug-ins or understanding the existing ones
Nmap (network mapper)	A free and open-source (license) utility for network discovery and security auditing. Many systems and network administrators find it useful for tasks such as network inventory, managing service upgrade schedules, and monitoring host or service uptime	Provides host information on the network, services (applications, version), OS/versions, packet filters/firewalls information, etc. Designed to scan large networks rapidly, but works on single host systems
OSSEC HIDS	Designed as an IDS	Performs log analysis, integrity checking, rootkit detection, time-based alerting, and active response. In addition to its IDS functionality, it is commonly used as a security event manager (SEM)/SIM solution. Because of its powerful log analysis engine, it can monitor and analyze firewalls, IDSs, Web servers, and authentication logs

(continued)

(continued)

Security tool	Function	Security information
Snort	A network intrusion detection and prevention system for traffic analysis and packet logging on IP networks	Detects thousands of worms, vulnerability exploit attempts, port scans, and other suspicious behavior
tcpdump	A network sniffer	Tracks network problems or monitoring activity
Wireshark™	An open-source, multi-platform network protocol analyzer. Formerly known as Ethereal®	Examines data from a live network or from a captured file on disk. It provides view on the reconstructed TCP session

Sectools.org [19]

Appendix G
Acronyms and Abbreviations

Term	Description
ARP	Address Resolution Protocol
ASCII	American Standard Code for Information Interchange
AV	Antivirus
BMP	Bitmap
CAPEC	Common Attack Pattern Enumeration and Classification
CS&C	Office of Cybersecurity and Communications
CSLP	Compressed Serial Line Protocol
CVRF	Common Vulnerability Reporting Framework
CybOX™	Cyber Observable eXpression
DCS	Distributed Control System
DDoS	Distributed Denial of Service
DFXML	Digital Forensics Extensible Markup Language
DHCP	Dynamic Host Configuration Protocol
DHS	U.S. Department of Homeland Security
DLC	Data Link Control
DMZ	Demilitarized Zone
DOJ	U.S. Department of Justice
DoS	Denial of Service
DVC	Digital Valve Controller
EBCDIC	Extended Binary Coded Decimal Interchange Code
ESD	Emergency Shutdown System
FBI	Federal Bureau of Investigation
FTP	File Transfer Protocol
GIF	Graphics Interchange Format
HITRAC	Homeland Infrastructure Threat and Risk Analysis Center

(continued)

© Springer Nature Switzerland AG 2021
Fiedelholtz, *The Cyber Security Network Guide*, Studies in Systems,
Decision and Control 274, https://doi.org/10.1007/978-3-030-61591-8

(continued)

Term	Description
HMI	Human-Machine Interface
I/O	Input/Output
IAA	Infrastructure Impact Assessments
IAB	Internet Architecture Board
IAC	Integrated Analysis Cell
ICANN	Internet Corporation for Assigned Names and Numbers
ICMP	Internet Control Message Protocol
ICS	Industrial Control System
ICS-CERT	Industrial Control System-Computer Emergency Readiness Team
IDS	Intrusion Detection System
IED	Intelligent Electronic Device
IEEE	Institute of Electrical and Electronics Engineers
IIS	Infrastructure Impact Summaries
IOC	Infrastructure of Concern
IP	Internet Protocol
IPS	Intrusion Protection System
IQL	Infrastructure Quick Look
iRAC	Incident Risk Analysis Cell
IRQ	In Response to your Question
ISAC	Information Sharing and Analysis Center
ISP	Internet Service Provider
IT	Information Technology
JPEG	Joint Photographic Experts Group
LAN	Local Area Network
LLC	Logical Link Control
MAC	Medium Access Control
MAEC	Malware Attribute Enumeration and Characterization
MITM	Man-in-the-Middle
MPEG	Motion Picture Experts Group
MS-ISAC	Multi-State Information Sharing and Analysis Center
MTU	Master Terminal Unit
NCC	National Coordinating Center for Telecommunications
NCCIC	National Cybersecurity and Communications Integration Center
NCSD	National Cyber Security Division
NFAT	Network Forensic Analysis Tool
NFS	Network File System
NIAC	National Infrastructure Advisory Council

(continued)

(continued)

Term	Description
NICC	National Infrastructure Coordinating Center
NIPP	National Infrastructure Protection Plan
NIST	National Institute of Standards and Technology
Nmap	Network Mapper
NOC	National Operations Center
NSA	National Security Agency
NSI	Nationwide Suspicious Activity Reporting Initiative
OS	Operating System
OSI	Open Systems Interconnection
OVAL	Open Vulnerability and Assessment Language
PC	Personal Computer
PCX	Personal Computer Exchange
PDA	Personal Digital Assistant
PHP	Personal Home Page
PLC	Programmable Logic Controller
POP	Point of Presence
PPP	Point-to-Point Protocol
RFA	Request for Analysis
RFI	Request for Information
RTU	Remote Terminal Unit
SAR	Suspicious Activity Report
SCADA	Supervisory Control and Data Acquisition
SCP	Session Control Protocol
SEM	Security Event Manager
SIEM	Security Information and Event Management
SIM	Security Information Management
SIS	Safety Instrumented System
SLIP	Serial Line Internet Protocol
SMS	Systems Management Server
SMTP	Simple Mail Transfer Protocol
SQL	Structured Query Language
SSH	Secure Shell Hash
SSL	Secure Socket Layer
STIX™	Structured Threat Information eXpression
TAXII™	Trusted Automated eXchange of Indicator Information
TCP	Transmission Control Protocol
TIFF	Tagged Image File Format

(continued)

(continued)

Term	Description
TLS	Transport Layer Security
TMH	TAXII™ Message Handler
TTA	TAXII™ Transfer Agent
TTP	Tactics, Techniques, and Procedures
UDP	User Datagram Protocol
UPS	Uninterruptible Power Supply
URL	Uniform Resource Locator
US-CERT	U.S.-Computer Emergency Readiness Team
UTM	Unified Threat Management
WAN	Wide Area Network
WLAN	Wireless Local Area Network
XML	Extensible Markup Language
ZIP	Zone Information Protocol

Appendix H
Glossary of Terms[a]

Term	Definition
Address Resolution Protocol (ARP)	A network layer protocol used to convert an IP address into a physical address (called a Data Link Control [DLC] address), such as an Ethernet address. A host wishing to obtain a physical address broadcasts an ARP request onto the TCP/IP network. The host on the network that has the IP address in the request then replies with its physical hardware address
American Standard Code for Information Interchange (ASCII)	ASCII is a code for representing English characters as numbers, with each letter assigned a number from 0 to 127. For example, the ASCII code for uppercase M is 77. Most computers use ASCII codes to represent text, which makes it possible to transfer data from one computer to another*
Antivirus (AV) tools	Software products and technology used to detect malicious code, prevent it from infecting a system, and remove malicious code that has infected the system
Attack	An attempt to gain unauthorized access to system services, resources, or information, or an attempt to compromise system integrity, availability, or confidentiality
Authentication	Verifying the identity of a user, process, or device, often as a prerequisite to allowing access to resources in an information system
Authorization	The right or a permission that is granted to a system entity to access a system resource
Backdoor	An undocumented way of gaining access to a computer system. A backdoor is a potential security risk

(continued)

© Springer Nature Switzerland AG 2021
Fiedelholtz, *The Cyber Security Network Guide*, Studies in Systems,
Decision and Control 274, https://doi.org/10.1007/978-3-030-61591-8

(continued)

Term	Definition
Buffer overflow	A condition at an interface under which more input can be placed into a buffer or data holding area than the capacity allocated, overwriting other information. Adversaries exploit such a condition to crash a system or to insert specially crafted code that allows them to gain control of the system
Common Attack Pattern Enumeration and Classification (CAPEC)	A publicly available, community-developed list of common attack patterns along with a comprehensive schema and classification taxonomy. Attack patterns are descriptions of common methods for exploiting software systems. They derive from the concept of design patterns applied in a destructive rather than constructive context and are generated from in-depth analysis of specific real-world exploit examples[b]
Compressed Serial Line Protocol (CSLP)	A compression version of SLIP. The protocol has no effect on the data payload of a packet and is independent of any compression used by the serial line modem
Confidentiality	Preserving authorized restrictions on information access and disclosure, including means for protecting personal privacy and proprietary information
Control center	An equipment structure or group of structures (facilities) from which a process is measured, controlled, and/or monitored
Control system	A system in which deliberate guidance or manipulation is used to achieve a prescribed value for a variable. Control systems include SCADA, DCS, PLCs and other types of industrial measurement and control systems
Cyber Observable eXpression (CybOX™)	A standardized schema for the specification, capture, characterization, and communication of events or stateful properties that are observable in the operational domain. A wide variety of high-level cybersecurity use cases rely on such information, including event management/logging, malware characterization, intrusion detection, incident response/management, and attack pattern characterization. CybOX™ provides a common mechanism (structure and content) for addressing cyber observables across and among this full range of use cases, improving consistency, efficiency, interoperability, and overall situational awareness[c]

(continued)

(continued)

Term	Definition
Database	A repository of information that usually holds plant-wide information, including process data, recipes, personnel data, and financial data
Digital Valve Controller (DVC)	Valves that use microprocessors to control process conditions such as flow, pressure, temperature, and liquid level by opening or closing in response to signals from controllers that compare a set point to a process variable whose value is provided by sensors that monitor changes in process conditions[d]
Distributed Control System (DCS)	In a control system, refers to control achieved by intelligence that is distributed about the process to be controlled, rather than by a centrally located single unit
Distributed Denial of Service (DDoS)	A type of denial of service (DoS) attack where multiple compromised systems—which are usually infected with a Trojan—are used to target a single system causing a DoS attack[e]
Domain Name System or Server (DNS)	An Internet service that translates domain names into IP addresses. Because domain names are alphabetic, they are easier to remember. The Internet, however, is based on IP addresses. Every time a domain name is used, therefore, a DNS service must translate the name into the corresponding IP address. For example, the domain name www.example.com might translate to 198.105.232.4. The DNS system is, in fact, its own network. If one DNS server does not know how to translate a particular domain name, it asks for another one, and so on, until the correct IP address is returned*
Dynamic Host Configuration Protocol (DHCP)	A protocol for assigning dynamic IP addresses to devices on a network. With dynamic addressing, a device can have a different IP address every time it connects to the network. In some systems, the device's IP address can even change while it is still connected. DHCP also supports a mix of static and dynamic IP addresses Dynamic addressing simplifies network administration because the software keeps track of IP addresses rather than requiring an administrator to manage the task. This means that a new computer can be added to a network without the hassle of manually assigning it a unique IP address. Many ISPs use dynamic IP addressing for dial-up users*

(continued)

(continued)

Term	Definition
Emergency Shutdown System (ESD)	Designed to minimize the consequences of emergency situations related to typically uncontrolled flooding, escape of hydrocarbons, or outbreak of fire in hydrocarbon carrying areas or areas that may otherwise be hazardous. An ESD system for a process control system includes an ESD valve and an associated valve actuator. An ESD controller provides output signals to the ESD valve in the event of a failure in the process control system. A solenoid valve responds to the ESD controller to vent the actuator to a fail state. A DVC test strokes the ESD valve. An impedance booster device enables the direct current powering of the solenoid valve and the DVC over a two-wire line while still permitting digital communication over the same two-wire line[f]
Encryption	Cryptographic transformation of data (called "plaintext") into a form (called "ciphertext") that conceals the data's original meaning to prevent it from being known or used. If the transformation is reversible, the corresponding reversal process is called "decryption," which is a transformation that restores encrypted data to its original state
Extended Binary Coded Decimal Interchange Code (EBCDIC)	An IBM code for representing characters as numbers. Although it is widely used on large IBM computers, most other computers, including PCs and Macintoshes, use ASCII codes*
Extensible Markup Language (XML)	A language developed especially for Web documents*
Fieldbus	A digital, serial, multi-drop, two-way data bus or communication path or link between low-level industrial field equipment such as sensors, transducers, actuators, local controllers, and even control-room devices. Use of fieldbus technologies eliminates the need for point-to-point wiring between the controller and each device. A protocol is used to define messages over the fieldbus network, with each message identifying a particular sensor on the network
Field Device	Equipment that is connected to the field side on an ICS. Types of field devices include RTUs, PLCs, actuators, sensors, HMIs, and associated communications

(continued)

(continued)

Term	Definition
File Transfer Protocol (FTP)	An Internet standard for transferring files over the Internet. FTP programs and utilities are used to upload and download Web pages, graphics, and other files between local media and a remote server that allows FTP access
Human-Machine Interface (HMI)	The hardware or software through which an operator interacts with a controller. An HMI can range from a physical control panel with buttons and indicator lights to an industrial PC with a color graphics display running dedicated HMI software
Identification	The process of verifying the identity of a user, process, or device, usually as a prerequisite for granting access to resources in an IT system
Information Technology (IT)	A term that refers to anything related to computing technology, such as networking, hardware, software, the Internet, or the people who work with these technologies[g]
Input/Output (I/O)	A general term for the equipment that is used to communicate with a computer as well as the data involved in the communications
Insider	An entity inside the security perimeter that is authorized to access system resources but uses them in a way not approved by those who granted the authorization
Integrity	Guarding against improper information modification or destruction, and includes ensuring information non-repudiation and authenticity
Intelligent Electronic Device (IED)	Any device incorporating one or more processors with the capability to receive or send data/control from or to an external source (e.g., electronic multifunction meters, digital relays, controllers)
Internet	The single interconnected world-wide system of commercial, government, educational, and other computer networks that share the set of protocols specified by the Internet Architecture Board (IAB) and the name and address spaces managed by the Internet Corporation for Assigned Names and Numbers (ICANN)

(continued)

(continued)

Term	Definition
Internet Control Message Protocol (ICMP)	The IP that allows the transfer of information from a computer system to another computer system using TCP/IP protocol. If there is a problem with the connection, error and status messages regarding the connection are sent using ICMP. ICMP can send back codes to a system explaining why a connection failed. These may be messages such as "Network unreachable" for a system that is down, or "Access denied" for a secure, password-protected system. ICMP may also provide routing suggestions to help bypass unresponsive systems. While ICMP can send a variety of different messages, most are never seen by the user. Even if the user does receive an error message, the software being used, such as a Web browser, has most likely already translated the message into simple language the user can understand[h]
Internet Protocol (IP)	Specifies the format of packets, also called datagrams, and the addressing scheme. Most networks combine IP with a higher-level protocol called Transmission Control Protocol (TCP), which establishes a virtual connection between a destination and a source. IP by itself is something like the postal system. It allows a sender to address a package and drop it in the system, but there is no direct link between the sender and the recipient. TCP/IP, on the other hand, establishes a connection between two hosts so that they can send messages back and forth for a period of time. The current version of IP is IPv4. A new version, called IPv6 or IPng, is under development*
Intrusion Detection System (IDS)	A security service that monitors and analyzes network or system events for the purpose of finding and providing real-time or near real-time warning of attempts to access system resources in an unauthorized manner
Intrusion Protection System (IPS)	A system that can detect an intrusive activity and can also attempt to stop the activity, ideally before it reaches its targets
Local Area Network (LAN)	A group of computers and other devices dispersed over a relatively limited area and connected by a communications link that enables any device to interact with any other on the network

(continued)

(continued)

Term	Definition
Logical Link Control (LLC)	A sublayer, one of two, that make up the Data Link Layer of the OSI Reference Model. The LLC layer controls frame synchronization, flow control, and error checking*
Malware	Software or firmware intended to perform an unauthorized process that will have adverse impact on the confidentiality, integrity, or availability of an information system. A virus, worm, Trojan horse, or other code-based entity that infects a host. Spyware and some forms of adware are also examples of malicious code (malware)
Malware Attribute Enumeration and Characterization (MAEC)	A standardized language for encoding and communicating high-fidelity information about malware based upon attributes such as behaviors, artifacts, and attack patterns. By eliminating the ambiguity and inaccuracy that exist in malware descriptions and by reducing reliance on signatures, MAEC aims to improve human-to-human, human-to-tool, tool-to-tool, and tool-to-human communication about malware; reduce potential duplication of malware analysis efforts by researchers; and allow for the faster development of countermeasures by enabling the ability to leverage responses to previously observed malware instances[i]
Man-in-the-Middle (MITM)	A type of active Internet attack in which the person attacking attempts to intercept, read, or alter information moving between two computers. MITM attacks are associated with 802.11 security, as well as with wired communication systems*
Master Terminal Unit (MTU)	The device that acts as the master in a SCADA system
Medium (or Media) Access Control (MAC)	A hardware address that uniquely identifies each node of a network. In Institute of Electrical and Electronics Engineers (IEEE) 802 networks, the DLC layer of the OSI Reference Model is divided into two sub-layers: the LLC layer and the MAC layer. The MAC layer interfaces directly with the network medium. Consequently, each different type of network medium requires a different MAC layer. On networks that do not conform to the IEEE 802 standards but do conform to the OSI Reference Model, the node address is called the DLC address*

(continued)

(continued)

Term	Definition
Motion Picture Experts Group (MPEG)	A working group of the International Organization for Standardization (ISO). The term also refers to the family of digital video compression standards and file formats developed by the group. MPEG generally produces better-quality video than competing formats, such as Video for Windows, Indeo, and QuickTime. MPEG files previously on personal computers (PCs) needed hardware decoders (codecs) for MPEG processing. Today, however, PCs can use software-only codecs including products from RealNetworks, QuickTime, or Windows Media Player*
Nationwide Suspicious Activity Reporting (NSI)	The Nationwide Suspicious Activity Reporting Initiative (NSI) is a collaborative effort led by the DOJ, Bureau of Justice Assistance, in partnership with DHS, the FBI, and State, local, tribal, and territorial law enforcement partners. This initiative provides law enforcement with another tool to help prevent terrorism and other related criminal activity by establishing a national capacity for gathering, documenting, processing, analyzing, and sharing suspicious activity reporting (SAR) information[j]
Network File System (NFS)	A client/server application designed by Sun Microsystems that allows all network users to access shared files stored on computers of different types. NFS provides access to shared files through an interface called the Virtual File System that runs on top of TCP/IP. Users can manipulate shared files as if they were stored locally on the user's own hard disk. With NFS, computers connected to a network operate as clients while accessing remote files, and as servers while providing remote users access to local shared files*
Network Forensic Analysis Tool (NFAT)	A collection of tools for network security surveillance, anomaly detection, analytics, and forensics. Examples of such tools are Nmap, SNORT, Wireshark, tcpdump, BackTrack, and ArcSight

(continued)

(continued)

Term	Definition
Network Mapper (Nmap)	A free and open-source (license) utility for network discovery and security auditing. Many systems and network administrators also find it useful for tasks such as network inventory, managing service upgrade schedules, and monitoring host or service uptime. Nmap uses raw IP packets in novel ways to determine what hosts are available on the network, what services (application name and version) those hosts are offering, what OS (and OS versions) they are running, what type of packet filters/firewalls are in use, and dozens of other characteristics. It was designed to rapidly scan large networks, but works well against single hosts. Nmap runs on all major computer OSs, and official binary packages are available for Linux, Windows, and Mac OSs[k]
Open Systems Interconnection (OSI)	An ISO standard for worldwide communications that defines a networking framework for implementing protocols in seven layers. Control is passed from one layer to the next, starting at the application layer in one station, proceeding to the bottom layer, over the channel to the next station and back up the hierarchy*
Open Vulnerability and Assessment Language (OVAL)	An information security community effort to standardize how to assess and report upon the machine state of computer systems. OVAL includes a language to encode system details, and an assortment of content repositories held throughout the community. Tools and services that use OVAL for the three steps of system assessment—representing system information, expressing specific machine states, and reporting the results of an assessment—provide enterprises with accurate, consistent, and actionable information so they may improve their security. Use of OVAL also provides for reliable and reproducible information assurance metrics and enables interoperability and automation among security tools and services[l]
Operating System (OS)	An integrated collection of service routines for supervising the sequencing of programs by a computer. An OS may perform the functions of input/output control, resource scheduling, and data management. It provides application programs with the fundamental commands for controlling the computer

(continued)

(continued)

Term	Definition
Password	A string of characters (letters, numbers, and other symbols) used to authenticate an identity or to verify access authorization
Personal Digital Assistant (PDA)	A handheld device that combines computing, telephone/fax, Internet, and networking features. A typical PDA can function as a cellular phone, fax sender, Web browser, and personal organizer*
Personal Home Page (PHP)	The server space provided by the web host to the subscribers for a fee.* PHP is an HTML-embedded Web scripting language. PHP also stands for hypertext preprocessor (a recursive acronym) is a code that can be inserted into the HTML of a Web page. When a PHP page is accessed, the PHP code is read or "parsed" by the server the page resides on. The output from the PHP functions on the page, are typically returned as HTML code which can be read by the browser. Because the PHP code is transformed into HTML before the page is loaded, users cannot view the PHP code on a page. This make PHP pages secure enough to access databases and other secure information[m]
Phishing	A method of tricking individuals into disclosing sensitive personal information by claiming to be a trustworthy entity in an electronic communication (e.g., Internet Web sites)
Port	The entry or exit point from a computer for connecting communications or peripheral devices
Port scanning	The process of using a program to remotely determine which ports on a system are open (e.g., whether systems allow connections through those ports)
Point-to-Point Protocol (PPP)	A method of connecting a computer to the Internet. PPP is more stable than the older SLIP protocol and provides error checking features. Working in the data link layer of the OSI Reference Model, PPP sends the computer's TCP/IP packets to a server that puts them onto the Internet*

(continued)

(continued)

Term	Definition
Programmable Logic Controller (PLC)	A solid-state control system that has a user-programmable memory for storing instructions for the purpose of implementing specific functions such as I/O control, logic, timing, counting, three-mode proportional-integral-derivative control, communication, arithmetic, and data and file processing
Protocol	A set of rules (i.e., formats and procedures) to implement and control some type of association (e.g., communication) between systems
Protocol analyzer	A device or software application that enables the user to analyze the performance of network data so as to ensure that the network and its associated hardware/software are operating within network specifications
Real-time	Pertaining to the performance of a computation during the actual time that the related physical process transpires so that the results of the computation can be used to guide the physical process
Remote access	Access by users (or information systems) communicating external to an information system security perimeter
Remote Terminal Unit (RTU)	A computer with radio interfacing used in remote situations where communications via wire is unavailable. Usually used to communicate with remote field equipment. PLCs with radio communication capabilities are also used in place of RTUs
Risk	The level of impact on agency operations (including mission, functions, image, or reputation), agency assets, or individuals, resulting from the operation of an information system, given the potential impact of a threat and the likelihood of that threat occurring
Risk assessment	The process of identifying risks to agency operations (including mission, functions, image, or reputation), agency assets, or individuals by determining the probability of occurrence, the resulting impact, and additional security controls that would mitigate this impact. Part of risk management, synonymous with risk analysis. Incorporates threat and vulnerability analyses

(continued)

(continued)

Term	Definition
Risk management	The process of managing risks to agency operations (including mission, functions, image, or reputation), agency assets, or individuals resulting from the operation of an information system. It includes risk assessment; cost-benefit analysis; the selection, implementation, and assessment of security controls; and the formal authorization to operate the system. The process considers effectiveness, efficiency, and constraints due to laws, directives, policies, or regulations
Router	A computer that is a gateway between two networks at OSI layer 3 and that relays and directs data packets through that inter-network. The most common form of router operates on IP packets
Safety Instrumented System (SIS)	A system that is composed of sensors, logic solvers, and final control elements whose purpose is to take the process to a safe state when predetermined conditions are violated. Other terms commonly used include EDS, safety shutdown system (SSD), and safety interlock system
Security Event Manager (SEM)	SEM provides the capability to process near real-time data from security devices and systems to determine when security events of interest have occurred[n]
Security Information and Event Management (SIEM)	The SIEM tool provides threat detection and security incident response through the real-time collection and historical analysis of security events from a wide variety of event and contextual data sources. It also supports compliance reporting and incident investigation through analysis of historical data from these sources. The core capabilities of SIEM technology are a broad scope of event collection and the ability to correlate and analyze events across disparate sources[o]

(continued)

(continued)

Term	Definition
Security Information Management (SIM)	A type of software that automates the collection of event log data from security devices, such as firewalls, proxy servers, IDSs, and AV software. The SIM translates the logged data into correlated and simplified formats. Many SIM architectures provide security reporting, analysis, and reporting for Sarbanes-Oxley, Basel II, the Health Insurance Portability and Accountability Act of 1996, the Federal Information Security Management Act of 2002, and Visa Cardholder Information Security Program compliance audits*
Sensor	A device that produces a voltage or current output that is representative of some physical property being measured (e.g., speed, temperature, flow)
Serial Line Internet Protocol (SLIP)	An IP for connection to the Internet via a dial-up connection. Developed in the 1980s when modem communications typically were limited to 2400 bits per second, it was designed for simple communication over serial lines. SLIP can be used on RS-232 serial ports and supports asynchronous links. A more common protocol is PPP because it is faster and more reliable and supports functions that SLIP does not, such as error detection, dynamic assignment of IP addresses, and data compression. In general, ISPs offer only one protocol, although some support both protocols*
Session Control Protocol (SCP)	A simple protocol that lets a server and client conduct multiple conversations over a single TCP connection. The protocol is designed to be simple to implement, and is modeled after TCP[p]
Simple Mail Transfer Protocol (SMTP)	A protocol for sending email messages between servers. Most email systems that send mail over the Internet use SMTP to send messages from one server to another; the messages can then be retrieved with an email client using either point of presence (POP) or Internet message access protocol (IMAP). In addition, SMTP is generally used to send messages from a mail client to a mail server. This is why the user needs to specify both the POP or IMAP server and the SMTP server when configuring an email application*

(continued)

(continued)

Term	Definition
Secure Shell Hash (SSH)	A program to log into another computer over a network, to execute commands in a remote machine, and to move files from one machine to another. It provides strong authentication and secure communications over insecure channels. SSH protects a network from attacks such as IP spoofing, IP source routing, and DNS spoofing. An attacker who has managed to take over a network can only force SSH to disconnect but cannot play back the traffic or hijack the connection when encryption is enabled. When using SSH's slogin (instead of rlogin) the entire login session, including transmission of password, is encrypted; therefore it is almost impossible for an outsider to collect passwords*
Secure Socket Layer (SSL)	A protocol developed by Netscape for transmitting private documents via the Internet. SSL uses a cryptographic system that uses two keys to encrypt data — a public key known to everyone and a private or secret key known only to the recipient of the message. Both Netscape Navigator and Internet Explorer support SSL, and many Web sites use the protocol to obtain confidential user information, such as credit card numbers. By convention, URLs that require an SSL connection start with *https* instead of *http**
Social engineering	An attempt to trick someone into revealing information (e.g., a password) that can be used to attack systems or networks
Spyware	Software that is secretly or surreptitiously installed onto an information system to gather information on individuals or organizations without their knowledge; a type of malicious code
Structured Query Language (SQL)	A standardized query language for requesting information from a database*
Structured Threat Information eXpression (STIX™)	The STIX™ standardized language intends to convey the full range of potential structured cyber threat information and strives to be fully expressive, flexible, extensible, automatable, and as human-readable as possible[q]
Stuxnet	A group of sophisticated malware worms that primarily target SCADA systems for large infrastructure assets (e.g., industrial power plants). The original Stuxnet worm was discovered in 2010, and numerous variations of it have been identified since then*

(continued)

(continued)

Term	Definition
Supervisory Control and Data Acquisition (SCADA)	A generic name for a computerized system that is capable of gathering and processing data and applying operational controls over long distances. Typical uses include power transmission and distribution and pipeline systems. SCADA was designed for the unique communication challenges (e.g., delays, data integrity) posed by the various media that must be used, such as phone lines, microwave, and satellite. Usually shared rather than dedicated
Suspicious Activity Report (SAR)	A report that provides the mechanism for anyone to report suspicious activity to a Federal agency such as DHS, local law enforcement, etc. This initiative provides law enforcement with another tool to help prevent terrorism and other terrorism-related crime by establishing a national capacity for gathering, documenting, processing, analyzing, and sharing SAR information in a report form[r]
Systems Management Server (SMS)	A set of Microsoft tools that helps manage PCs connected to a LAN. SMS enables network administrator to inventory all hardware and software on the network and to store it in a database. Using this database, SMS can then perform software distribution and installation over the LAN*
Tagged Image File Format (TIFF)	One of the most widely supported file formats for storing bitmapped images on personal computers (both PCs and Macintosh computers). Other popular formats are bitmap (BMP)and Personal Computer Exchange (PCX). TIFF graphics can be any resolution, and they can be black and white, gray-scaled, or color. Files in TIFF format often end with a .tif extension*
Technical controls	The security controls (i.e., safeguards or countermeasures) for an information system that are primarily implemented and executed by the information system through mechanisms contained in the hardware, software, or firmware components of the system
Threat	Any circumstance or event with the potential to adversely affect agency operations (including mission, functions, image, or reputation), agency assets, or individuals through an information system via unauthorized access, destruction, disclosure, modification of information, and/or DDoS

(continued)

(continued)

Term	Definition
Transmission Control Protocol (TCP)	One of the main protocols in TCP/IP networks. Whereas the IP protocol deals only with packets, TCP enables two hosts to establish a connection and exchange streams of data. TCP guarantees delivery of data and guarantees that packets will be delivered in the same order in which they were sent
Transport Layer Security (TLS)	A protocol that guarantees privacy and data integrity between client/server applications communicating over the Internet*
Trojan Horse	A computer program that appears to have a useful function, but also has a hidden and potentially malicious function that evades security mechanisms, sometimes by exploiting legitimate authorizations of a system entity that invokes the program
Trusted Automated eXchange of Indicator Information (TAXII™)	Defines a set of services and message exchanges that, when implemented, enable sharing of actionable cyber threat information across organization and product/service boundaries. TAXII™, through its member specifications, defines concepts, protocols, and message exchanges to exchange cyber threat information for the detection, prevention, and mitigation of cyber threats[s]
Unauthorized access	A circumstance in which a person gains logical or physical access without permission to a network, system, application, data, or other resource
Unified threat Management (UTM)	A term first used by Internet Data Corporation to describe a category of security appliances which integrates a range of security features into a single appliance. UTM appliances combine firewall, gateway AV, and intrusion detection and prevention capabilities into a single platform. UTM is designed to protect users from blended threats while reducing complexity*
Uniform Resource Locator (URL)	The global address of documents and other resources on the Internet. The first part of the URL, the protocol identifier, indicates what protocol to use; the second part, the resource name, specifies the IP address or the domain name where the resource is located. These two elements are separated by a colon and two forward slashes*

(continued)

(continued)

Term	Definition
Uninterruptible Power Supply (UPS)	Also known as uninterruptible power source, UPS, or battery/flywheel backup, it is an electrical apparatus that provides emergency power to a load when the input power source, typically the utility mains, fails[t]
User Datagram Protocol (UDP)	A connectionless protocol that, like TCP, runs on top of IP networks. Unlike TCP/IP, UDP/IP provides very few error recovery services, offering instead a direct way to send and receive datagrams over an IP network. It is used primarily for broadcasting messages over a network*
Virus	A hidden, self-replicating section of computer software, usually malicious logic, that propagates by infecting (i.e., inserting a copy of itself into and becoming part of) another program. A virus cannot run by itself; it requires that its host program be run to make the virus active
Vulnerability	Weakness in an information system, system security procedures, internal controls, or implementation that could be exploited or triggered by a threat source
Wide Area Network (WAN)	A physical or logical network that provides data communications to a larger number of independent users than are usually served by a LAN and that is usually spread over a larger geographic area than that of a LAN
Wireless Local Area Network (WLAN)	A LAN that uses high-frequency radio signals to transmit and receive data over distances of a few hundred feet; uses Ethernet protocol[u]
Worm	A computer program that can run independently, can propagate a complete working version of itself onto other hosts on a network, and may consume computer resources destructively
Zone Information Protocol (ZIP)	Provides applications and processes with access to zone names. A zone is a logical grouping of nodes in an AppleTalk Internet, and each zone is identified by a name. A zone name is typically used to identify an affiliation between a group of nodes, such as a group of nodes belonging to a particular department within an organization[v]

[a]Stouffer et al. [20]
[b]The MITRE Corporation [21]
[c]The MITRE Corporation [22]
[d]Wikipedia.org [23]
[e]Quinstreet Inc. (Webopedia) [24]
[f]Esoteric Automation & Control Technologies [25]
[g]TechTerms.com [26]

[h]TechTerms.com [27]
[i]The MITRE Corporation [28]
[j]Institute for Intergovernmental Research [29]
[k]Nmap.org [30]
[l]The MITRE Corporation [31]
[m]TechTerms.com [32]
[n]Gartner [33]
[o]Ibid.
[p]W3.org Simon Spero, UNC Sunsite/EIT [34]
[q]The MITRE Corporation [35]
[r]Federal [36]
[s]The MITRE Corporation [8]
[t]Farlax Inc. (Freedictionary.com) [37]
[u]Farlax Inc. [38]
[v]Apple.com [39]

Note Definitions not footnoted are from NIST SP800-82 standard; definitions marked with an asterisk (*) are from Webopedia: http://www.webopedia.com

References

1. *Forbes*, Top 10 Security Issues That Will Destroy Your Computer in 2013. Available at http://www.forbes.com/sites/kenrapoza/2012/12/05/top-10-security-issues-that-will-destroy-your-computer-in-2013 (2013). Accessed 25 Mar 2013
2. Croucher, M.: Trojan attack may not be limited to Iran. The National (2012, April 27). Available at http://www.thenational.ae/news/uae-news/trojan-attack-may-not-be-limited-to-iran. Accessed 25 Mar 2013
3. The MITRE Corporation: Standardizing Cyber Threat Intelligence Information with the Structured Threat Information eXpression (STIX™). Available at http://stix.mitre.org/about/documents/STIX_Whitepaper_v1.0.pdf (2013b). Accessed 25 Mar 2013
4. Yavvari, C., Tokhtabayev, A., Rangwala, H., Stavrou, A.: Malware Characterization using Behavioral Components. George Mason University. Available at http://cs.gmu.edu/~astavrou/research/Behavioral_Map.pdf (2012). Accessed 25 Mar 2013
5. Butler, J.M.: Benchmarking Security Information Event Management (SIEM). SANS Institute (2009, February). Available at http://www.sans.org/reading_room/analysts_program/eventMgt_Feb09.pdf. Accessed 25 Mar 2013
6. SANS Institute; Critical Control 14: Maintenance, Monitoring, and Analysis of Audit Logs. Available at http://www.sans.org/critical-security-controls/control.php?id=14 (2013). Accessed 25 Mar 2013
7. The MITRE Corporation, Cyber Observable eXpression—CybOX: A Structured Language for Cyber Observables. Available at http://measurablesecurity.mitre.org/docs/cybox-intro-handout.pdf (undated). Accessed 25 Mar 2013
8. The MITRE Corporation: TAXII—Trusted Automated eXchange of Indicator Information. Available at http://taxii.mitre.org (2013). Accessed 25 Mar 2013
9. Davidson, M., Schmidt, C.: TAXII Overview Version 1.0. The MITRE Corporation. Available at http://taxii.mitre.org/specifications/version1.0/TAXII_Overview.pdf (2013). Accessed 25 Mar 2013
10. University of Washington: The OSI Model. Available at http://www.washington.edu/lst/help/computing_fundamentals/networking/osi (2011). Accessed 25 Mar 2013
11. University of Oregon: Security Vulnerabilities: Up and Down the OSI Stack. Available at http://pages.uoregon.edu/joe/nitrd/up-and-down-the-osi-model.html (undated). Accessed 16 Apr 2013
12. TechExams.net: Thread: Security OSI Model. Available at http://www.techexams.net/forums/ec-council-ceh-chfi/65437-security-osi-model.html (2002-2011). Accessed 16 Apr 2013
13. About.com: IP Tutorial—Subnet Masks and Subnetting. Available at http://compnetworking.about.com/od/workingwithipaddresses/l/aa043000b.htm (2013). Accessed 25 Mar 2013
14. Wireshark.org: Chapter 6. Working with Captured Packets. Available at http://www.wireshark.org/docs/wsug_html_chunked/ChapterWork.html (undated). Accessed 25 Mar 2013

15. SANS Institute: ICMP Attacks Illustrated. Available at http://www.sans.org/reading_room/whi tepapers/threats/icmp-attacks-illustrated_477 (undated). Accessed 25 Mar 2013
16. OWASP (Open Web Application Security Project): Session Timeout. Available at https://www. owasp.org/index.php/Session_Timeout (2012). Accessed 25 Mar 2013
17. Microsoft: The OSI Model's Seven Layers Defined and Functions Explained. Available at http://support.microsoft.com/kb/103884 (2013). Accessed 25 Mar 2013
18. TutorialsWeb.com: The OSI (Open Systems Interconnection) model. Available at http://www. tutorialsweb.com/networking/osi-model-application-layer.htm (2003–2011). Accessed 25 Mar 2013
19. Sectools.org: SecTools.org: Top 125 Network Security Tools. Available at http://sectools.org (2013). Accessed 25 Mar 2013
20. Stouffer, K., Falco, J., Scarfone, K.: Special publication SP800-82: guide to industrial control systems (ICS) security. NIST (2011). Available at http://csrc.nist.gov/publications/nistpubs/ 800-82/SP800-82-final.pdf (2011). Accessed 25 Mar 2013
21. The MITRE Corporation: CAPEC—Common Attack Pattern Enumeration and Classification. Available at http://capec.mitre.org/index.html (2013). Accessed 25 Mar 2013
22. The MITRE Corporation: CybOX—Cyber Observable eXpression. Available at http://cybox. mitre.org (2013). Accessed 25 Mar 2013
23. Wikipedia.org: Control Valves. Available at http://en.wikipedia.org/wiki/Control_valves (2013). Accessed 25 Mar 2013
24. Quinstreet Inc. (Webopedia): DDoS Attack—Distributed Denial of Service. Available at http:// www.webopedia.com/TERM/D/DDoS_attack.html (2013). Accessed 25 Mar 2013
25. Esoteric Automation & Control Technologies: Emergency Shut Down (ESD) System. Available at http://www.esotericautomation.com/Emergency-Shut-Down-System.html (undated). Accessed 25 Mar 2013
26. TechTerms.com: IT. Available at http://www.techterms.com/definition/it (undated a). Accessed 25 Mar 2013
27. TechTerms.com: ICMP. Available at http://www.techterms.com/definition/icmp (undated b). Accessed 25 Mar 2013
28. The MITRE Corporation: MAEC—Malware Attribute Enumeration and Characterization. Available at http://maec.mitre.org (2013). Accessed 25 Mar 2013
29. Institute for Intergovernmental Research: The Nationwide SAR Initiative. Available at http:// nsi.ncirc.gov (2013). Accessed 25 Mar 2013
30. Nmap.org: Introduction. Available at http://nmap.org (undated). Accessed 25 Mar 2013
31. The MITRE Corporation: OVAL—Open Vulnerability and Assessment Language. Available at https://oval.mitre.org (2013). Accessed 25 Mar 2013
32. TechTerms.com: PHP. Available at http://www.techterms.com/definition/php (2013). Accessed 25 Mar 2013
33. Gartner: Security information and Event Management (SIEM). Available at http://www.gar tner.com/it-glossary/?s=security+event+manager (2013). Accessed 25 Mar 2013
34. W3.org Simon Spero, UNC Sunsite/EIT: Session Control Protocol. Available at http://www. w3.org/Protocols/HTTP-NG/http-ng-scp.html (undated). Accessed 25 Mar 2013
35. The MITRE Corporation: STIX—Structured Threat Information eXpression. Available at http://stix.mitre.org (2013). Accessed 25 Mar 2013
36. Federal: Report Suspicious Activity. Available at http://www.dhs.gov/how-do-i/report-suspic ious-activity (undated d). Accessed 25 Mar 2013
37. Farlax Inc. (Freedictionary.com): Uninterruptible power supply. Available at http://encyclope dia.thefreedictionary.com/UPS+(uninterruptible+power+source) (2013a). Accessed 25 Mar 2013
38. Farlax Inc.: Wireless local area network. Available at http://www.thefreedictionary.com/wir eless+local+area+network (2013b). Accessed 25 Mar 2013
39. Apple.com: Chapter 4: Zone Information Protocol (ZIP). Available at http://developer. apple.com/legacy/library/documentation/mac/pdf/Networking/ZIP.pdf (1995). Accessed 25 Mar 2013

Uncited References

40. About.com: DMZ—Demilitarized Zone. Available at http://compnetworking.about.com/cs/net worksecurity/g/bldef_dmz.htm (undated). Accessed 25 Mar 2013
41. Copeland, G.: Infrastructure dependencies: the big rocks panel—private sector perspective. CSC (2011). Available at http://www.dtic.mil/ndia/2011DIBCIP/Copeland.pdf (2011). Accessed 25 Mar 2013
42. Federal: Privacy Compliance Review of the EINSTEIN Program. Available at http://www.dhs. gov/xlibrary/assets/privacy/privacy_privcomrev_nppd_ein.pdf (2012). Accessed 25 Mar 2013
43. Federal: Bottom-Up Review Report. Available at http://www.dhs.gov/xlibrary/assets/bur_bot tom_up_review.pdf (2010). Accessed 14 Aug 2013
44. Federal: About the National Cybersecurity and Communications Integration Center. Available at http://www.dhs.gov/about-national-cybersecurity-communications-integration-center (undated a). Accessed 22 Mar 2013
45. Federal: National Coordinating Center for Telecommunications. Available at http://www.dhs. gov/national-coordnating-center-telecommunications (undated b). Accessed 25 Mar 2013
46. Fedreal: Chemical Sector. Available at http://www.dhs.gov/chemical-sector (undated c). Accessed 25 Mar 2013
47. Federal, CSSP NCSD: Cyber Security Assessments of Industrial Control Systems. Available at http://ics-cert.us-cert.gov/sites/default/files/Cyber_Security_Assessments_of_Industrial_C ontrol_Systems.pdf (2010). Accessed 25 Mar 2013
48. Federal, HITRAC IAC: Integrated Analysis Cell—Standard Operating Procedures (2013)
49. Federal, ICS-CERT: Figure 1: Communications access to control systems. Available at https:// ics-cert.us-cert.gov/sites/default/files/transimages/figure1.jpg (2013). Accessed 25 Mar 2013
50. GPO (U.S. Government Printing Office): Examining the Cyber Threat to Critical Infrastructure and the American Economy. Available at http://www.gpo.gov/fdsys/pkg/CHRG-112hhrg72 221/pdf/CHRG-112hhrg72221.pdf (2012). Accessed 25 Mar 2013
51. HKSAR (The Government of the Hong Kong Special Administrative Region): An Overview of Vulnerability Scanners. Available at http:/www.infosec.gov.hk/english/technical/files/vulner ability.pdf (2008, February). Accessed 25 Mar 2013
52. The MITRE Corporation: Use Cases. Available at http://stix.mitre.org/language/usecases.html (2013). Accessed 25 Mar 2013
53. The MITRE Corporation: About CybOX (2012, January 17). Available at http://cybox.mitre. org/about. Accessed 16 Apr 2013
54. The MITRE Corporation: Cyber Observable eXpression—CybOX: A Structured Language for Cyber Observables. Available at http://measurablesecurity.mitre.org/docs/cybox-intro-han dout.pdf (undated). Accessed 25 Mar 2013
55. NIAC: Critical Infrastructure Resilience—Final Report and Recommendations (2009, September 8). Available at http://www.dhs.gov/xlibrary/assets/niac/niac_critical_infrastruc ture_resilience.pdf. Accessed 25 Mar 2013
56. Peerenboom, J.: Infrastructure Interdependencies: Overview of Concepts and Terminology. Argonne National Laboratory. Available at http://we-partner.org/onionbelt/wp-content/upl oads/2011/06/peerenboom_pdf.pdf (2001). Accessed 25 Mar 2013
57. Shell Global Solutions, UK: Safety instrumented systems. Available at http://www.idc-onl ine.com/technical_references/pdfs/instrumentation/safety_instrumented_systems.pdf (2003). Accessed 25 Mar 2013
58. Stoneburner, G., Goguen, A., Feringa, A.: Special publication 800-30—risk management guide for information technology systems. NIST (2013, July). Available at http://csrc.nist.gov/pub lications/nistpubs/800-30/sp800-30.pdf. Accessed 25 Mar 2013

Printed in the United States
by Baker & Taylor Publisher Services